Plant Genetic Resources and Climate Change

Plant Genetic Resources and Climate Change

Editors

Suresh Dhumal, Harshad Khuspe

scitus academics

Plant Genetic Resources and Climate Change

Edited by **Suresh Dhumal, Harshad Khuspe**

ISBN: 978-1-68117-086-2

Library of Congress Control Number: 2015935485

www.scitusacademics.com
Box No. 4766, 616 Corporate Way,
Suite 2, Valley Cottage,
NY 10989

Printed in United States of America on Acid Free Paper ♾

Contents

Preface

There is much less scepticism today about greenhouse gas-induced climate change and what its consequences might be, even though the full impacts of climate change cannot yet be predicted with certainty. On the other hand, the nature of weather variability – particularly in the northern hemisphere in recent years – has left some again questioning whether our climate really is warming. But the evidence is there for all to see, even as the sceptics refuse to accept the empirical data of increases in atmospheric CO2, for example, or the unprecedented summer melting of sea ice in the Arctic and the retreat of glaciers in the Alps. Over the past decade the world has experienced a number of severe climate events – wake-up calls to what might be the normal pattern in the future under a changed climate – such as extreme drought in one region, or unprecedented flooding in another. Even the 'normal' weather patterns of Western Europe appear to have become disrupted in recent years leading to increased stresses on agriculture. In the book, some of the latest perspectives about how genetic resources can contribute to achieving food security under the challenge of a changing climate. We also wanted to highlight some key issues for plant genetic resources management, to demonstrate how perspectives have changed over two decades, and discuss some of the actual responses and developments.

Editor

Chapter 1

Separating the Wheat from the Chaff – A Strategy to Utilize Plant Genetic Resources from Ex Situgenebanks

Jens Keilwagen[1,2], Benjamin Kilian[3], Hakan Özkan[4], Steve Babben[5], Dragan Perovic, Klaus F. X. Mayer[6], Alexander Walther[7], C. Hart Poskar[8], Frank Ordon[5], Kellye Eversole[9], Andreas Börner[3],Martin Ganal[10], Helmut Knüpffer[3], Andreas Graner[3], and Swetlana Friedel[1]

[1]Molecular Genetics, Leibniz Institute of Plant Genetics and Crop Plant Research (IPK), D-06466 Seeland OT Gatersleben, Germany,

[2]Institute for Biosafety in Plant Biotechnology, Julius Ku¨hn-Institut (JKI) - Federal Research Centre for Cultivated Plants, D-06484 Quedlinburg, Germany,

[3]Genebank, Leibniz Institute of Plant Genetics and Crop Plant Research (IPK), D-06466 Seeland OT Gatersleben, Germany,

[4]University of *Çukurova*, Faculty of Agriculture, Department of Field Crops, 01330 Adana, Turkey,

[5]Institute for Resistance Research and Stress Tolerance, Julius Ku¨hn-Institut (JKI) - Federal Research Centre for Cultivated Plants, D-06484

Quedlinburg, Germany,

[6]MIPS/Institute for Bioinformatics and Systems Biology, Helmholtz Center Munich, D-85764 Neuherberg, Germany,

[7]University of Gothenburg, Department of Earth Sciences, SE-405 30 Go¨teborg, Sweden,

[8]Physiology and Cell Biology, Leibniz Institute of Plant Genetics and Crop Plant Research (IPK), D-06466 Seeland OT Gatersleben, Germany,

[9]International Wheat Genome Sequencing Consortium, Eversole Associates, MD 20816 Bethesda, USA,

[10]TraitGenetics GmbH, Am Schwabeplan 1b, D-06466 Seeland OT Gatersleben, and Germany.

ABSTRACT

The need for higher yielding and better-adapted crop plants for feeding the world's rapidly growing population has raised the question of how to systematically utilize large genebank collections with their wide range of largely untouched genetic diversity. Phenotypic data that has been recorded for decades during various rounds of seed multiplication provides a rich source of information. Their usefulness has remained limited though, due to various biases induced by conservation management over time or changing environmental conditions. Here, we present a powerful procedure that permits an unbiased trait-based selection of plant samples based on such phenotypic data. Applying this technique to the wheat collection of one of the largest genebanks worldwide, we identified groups of plant samples displdiscovery paying contrasting phenotypes for selected traits. As a proof of concept for our ipeline, we resequenced the entire major but conserved flowering time locus *Ppd-D1* in just a few such selected wheat samples – and nearly doubled the number of hitherto known alleles.

Subject terms:

INTRODUCTION

Climate change and the rapidly growing human population trigger the demand for significantly improved crop plants[1, 2, 3, 4, 5, 6, 7]. At present, new plant varieties are mainly developed through the reshuffling of alleles present in the elite gene pool (crossing the best – hoping for the best) resulting in a more or less constant repertoire of alleles or even an erosion of genetic diversity. As a result, genetic gains have gradually slowed for most of the major crop species including wheat[8,9]. As a result, the relative increase in grain yield of wheat (*Triticum aestivum* L.), the most dominant crop species with a global acreage of 217 million ha (http://faostat.fao.org/), has fallen short of the global population growth rate in many parts of the world.

One of the most promising approaches to cope with this challenge is to valorize the rich genetic diversity conserved in large germplasm collections stored *ex situ* in genebanks for trait improvement by the introduction of novel alleles[10, 11, 12, 13]. However, the identification of plant samples (plant samples commonly referred to as 'accessions' in a genebank context) with specific combinations of traits from genebank collections is inherently difficult, because of varying regeneration intervals (due to improved storage conditions) and owing to the fact that collections were not regenerated in a systematic scheme (i.e. individual regeneration cycles did not comprise the same set of accessions). Furthermore, changes in conservation management over time (evolution of agricultural practices, equipment, storage and regeneration strategies, and genebank standards) impinge on the usefulness of phenotypic observations. Finally, environmental effects including inter-annual weather variability and long-term climate change affect plant phenology[4, 14,15, 16, 17, 18]. This is reflected by a shift of up to three weeks towards earlier flowering within the last 60 years, correlated with a significant increase of average spring temperatures over this period As a result, technical and environmental effects lead to highly heterogeneous phenotypic data which makes valorization of genetic resources a daunting task.

RESULTS AND DISCUSSION

The Normalized Rank Product is Less Biased than Other Descriptive Statistics

Against this backdrop, we aimed at exploiting long-term phenotypic data to enable systematic access to large genebank collections for upcoming genome resequencing and genome-wide association studies (Fig. 1). We propose a statistical approach, which allows for the consolidation of data sets that were collected over sixty years of seed multiplication. The approach aims at the comparison of accessions grown in different years using a standardized scale. This is achieved by calculating the Normalized Rank Product (NRP) to order accessions for a specific trait under consideration. Thus, accessions grown across multiple decades can be compared for particular traits using a common scale ranging from 0 to 1.

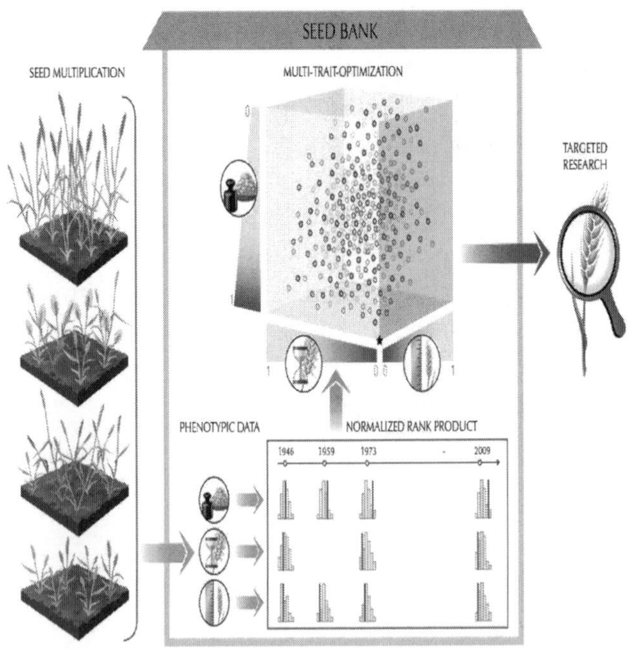

Figure 1: A novel strategy for valorizing genetic diversity stored in genebanks.

Every year genebank curators select accessions for seed multiplication. These accessions are grown in field trials and their phenotypic observations are recorded during the growing period. Annual ranking of these data is the first step in NRP analysis which allows comparing accessions that have been cultivated in different years and under different conditions. This example visualizes a wheat accession cultivated four times since 1946. The histograms indicate the distributions for all accessions cultivated in a given year. Missing histograms indicate missing phenotypic observations. Based on NRPs, MTO allows identifying accessions with specific combinations of traits that can be utilized for targeted plant research and breeding. The black asterisk and red dot in the cube represent the best virtual and one real accession, respectively, that simultaneously have early flowering time, small plant height and high thousand grain weight. In the histograms of absolute trait values, the actual accession is indicated by red lines. This figure has been generated using Adobe Photoshop CC and Adobe Illustrator CC.

For validation, the NRP was compared against descriptive statistics like the mean, for ranking accessions based on historical data for flowering time. Analyzing this trait by using a 9-year sliding window for the annual median flowering time (Fig. 2A), it becomes obvious that the flowering time depends on the year of cultivation. More specifically, early flowering was observed predominantly in recent decades (after 1980), while late flowering was observed in early decades. Based on this observation, we compare a varying number of accessions with best mean and NRP values by examining the percentage of accessions that have been cultivated for the first time after 1980 (Fig. 2B). For the mean, the percentage initially is very high and converges with increasing number of selected accessions to the percentage of accessions entered into the collection after 1980, whereas the percentage fluctuates around the global value for the NRP. This observation indicates that the NRP is less influenced by the shift of flowering time.

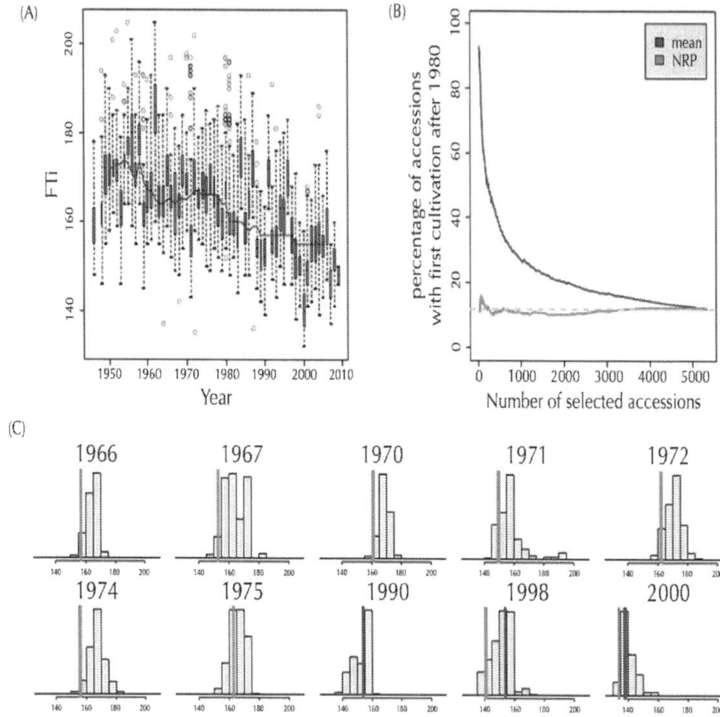

Figure 2: Validation of the NRP for flowering time of winter wheat.

(A) Box plots and trend of flowering time (FTi, in days of year, ordinate) for winter wheat between 1946 and 2010 (abscissa). The trend in blue is the 9-year sliding window for annual median flowering time indicating a clear shift towards earlier flowering. (B) Comparison of NRP with the naïve method of averaging the phenotypic observations (mean). The panel plots the percentage of accessions that have been cultivated for the first time after 1980 against the number of selected accessions. For the mean, the percentage is initially very high and converges to the global value. In contrast, the behavior of the NRP is almost constant. (C) Specific example of early flowering time for two winter wheat accessions. Green visualizes the accession TRI 7594 selected by NRP, while red visualizes the accession TRI 16575 selected by mean. Hence, the red accession has a smaller mean (149 < 153) but a higher NRP (0.51 > 0.06) than the green accession. However, the two common cultivations in 1998 and 2000 illustrate that the green accession typically flowers earlier than the red one. Plots were created with R (http://r-project.org).

To further validate the power of the NRP, two winter wheat accessions showing contradicting mean and NRP were scrutinized as an example. A direct comparison was facilitated because of data from 2 years of common cultivation (1998, 2000). In both years, the accession with smaller NRP flowered earlier than the one with smaller mean flowering time (Fig. 2C) again confirming the accuracy of the NRP. Finally, mean and NRP were assessed on all pairs of accessions that have been cultivated together in at least two years as described in the specific example above. Based on this evaluation, the NRP was shown to be less biased than the mean Thus, the NRP is a robust trait characteristics and can be used for instance for clustering and principal component analysis. As an example, every accession in the collection can be described by more than one normalized trait.

Multi-Trait Optimization Identifies Promising Accessions for Further Evaluation

In Fig. 1, the results are presented for three traits, namely, thousand grain weight, flowering time, and plant height. In general if one is interested in *n* different traits, one might span an *n*-dimensional hypercube. For the identification of promising accessions for further evaluation, we used a multi-trait optimization (MTO) approach based on the NRP to overcome an intrinsic structure caused by correlations between normalized traits and dependencies on passport data.

This approach selects accessions which are located near the corners of the cube corresponding to the most extreme trait combinations. Hence, opposing corners represent contrasting phenotypes. For example, the corner (1,0,0) indicated by an asterisk in Fig. 1 refers to the combination of high thousand grain weight, early flowering, and short plant height, whereas the corner (0,1,1) refers to the contrasting combination of low thousand grain weight, late flowering, and tall plant height. Based on that, MTO allows for identifying accessions breaking the existing correlations between traits, i.e., having combinations of traits that only rarely occur in a given gene pool (cf. empty corners in Fig. 3A,

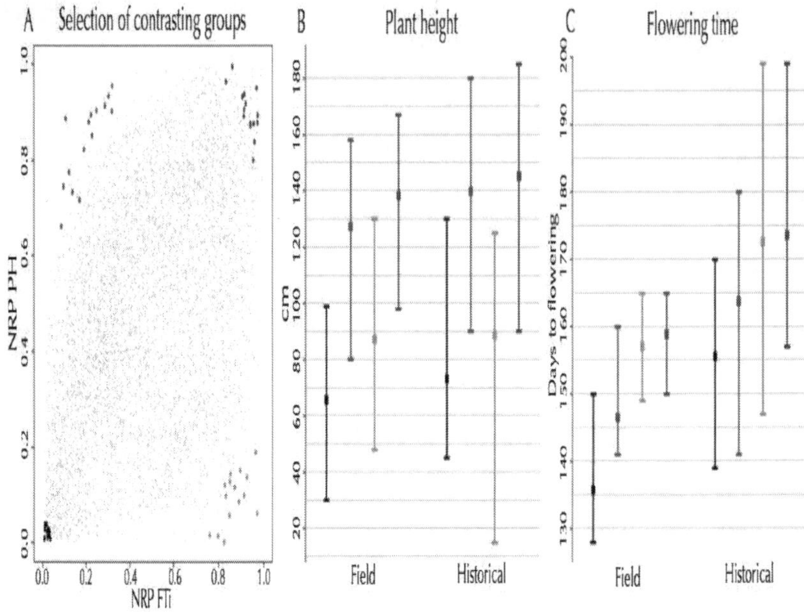

Figure 3: Validation of MTO for a wheat collection comprising 6,959 accessions

(A) depicts the results of the MTO for the normalized traits flowering time (NRP FTi) and plant height (NRP PH) selecting four contrasting groups with 15 accessions each. (B) and (C) compare the phenotyping results (minimum, mean and maximum of pre-normalized values) of the field experiment 2010/2011 and the historical phenotypic data (1946–2009) for plant height and flowering time, respectively. In all three panels, the colors encode the selected contrasting groups, where black represents short and early flowering accessions, red represents tall and early flowering, green represents short and late flowering, and blue represents tall and late flowering

To validate this MTO, we used long-term phenotypic data that were available for nearly 7,000 winter wheat accessions. For further validation, four groups of contrasting phenotypes were selected for extreme plant height and extreme flowering time and grown in a field experiment. In total, 60 accessions (15 accessions per contrasting group) were grown together, which allowed for the direct comparison of their phenotypes. Comparison with the legacy data revealed highly similar patterns demonstrating the efficacy of MTO

Proof of Concept: Resequencing at *Ppd-D1* Locus Nearly Doubled the Number of Known Alleles

To investigate the potential of trait-based selection for discovering new alleles, all 60 contrasting accessions obtained from MTO (Fig. 3A), represented by 96 individual plants, were resequenced at the major photoperiod response locus *Ppd-D1* For this, we utilized sequence information from the International Wheat Genome Sequencing Consortium (IWGSC)[19,20, 21]. Despite the small sample size and the high degree of conservation at this locus, three novel alleles were identified, nearly doubling the number of haplotypes known for bread wheat at this locus[22, 23, 24, 25] Two of the polymorphisms were located in coding regions upstream of the CCT domain leading to premature stop codons and thus most likely cause a loss of function.

In conclusion, we proposed a procedure to compare results from "non orthogonal" field trials which are a hallmark of the continuous reproduction schemes of genebank collections. The solution, illustrated here for wheat, may be applied to other crops and their wild relatives, making this strategy a standard approach to select for 'wanted' phenotypic combinations and to 'widen' the breeding bottleneck. This study demonstrates that the enormous amount of phenotypic data, recorded in genebanks over decades, can indeed be used to select accessions with desired combinations of traits. Computational methods as the NRP can be applied immediately and at almost zero cost - harvesting the investment of time, energy and money made in genebanks over generations and proving their continued value.

Based on NRPs and MTO of long-term phenotypic data, it is feasible to accurately assess the genotypic diversity and identify divergent accessions from comprehensive collections of any size, thereby avoiding the limitations entailed by the use of "core collections"[26]. Additional information including genotypic and pedigree information, if available, might be used to even better assess the genotypic diversity. Identification of a small set of contrasting phenotypes for flowering time immediately allowed for tapping into a substantial amount of novel allelic diversity at a major flowering time locus in wheat, which can be characterized and validated in future studies. Upcoming genome sequences for major cereal crops and new techniques including whole genome exome capture[19, 20, 21, 27] will allow for a much more extensive association of phenotypic and allelic diversity and thus help to bridge the genotype to phenotype gap.

METHODS

Federal Gene bank of Germany

The Federal *ex situ* Genebank for Agricultural and Horticultural Plant Species of Germany maintained at the Leibniz Institute of Plant Genetics and Crop Plant Research (IPK) in Gatersleben hosts one of the largest barley and wheat germplasm collections in the world[28]. The collection was established at its present location in 1945, and seed multiplication has been continuously performed since 1946.

The cultivation for seed multiplication has taken place every two to three years until 1976. After a cold storage facility was established in 1976, the average frequency of seed multiplication dropped drastically to every 10–30 years and accessions have been selected for seed multiplication if either the germination rate or the amount of available seeds dropped below a critical threshold[29].

Characterization and Evaluation Data

For IPK's seed bank, Characterization and Evaluation (C & E) data comprises dates of various growth stages (e.g., sowing, emergence, heading, flowering, ripening, harvest), winter survival, lodging, infection scores for some of the most important fungal diseases, plant height, and thousand-grain weight. In this paper, we focused the analysis on three key traits, namely, days to flowering (FTi), plant height (PH), and thousand-grain weight (TGW) for wheat (*Triticum aestivum*L.) and barley (*Hordeum vulgare* L.). Spring and winter types (growth habit) are distinguished for both cereals (Note: in Central and Northern Europe, winter types are sown in autumn - September until December -, while spring types are sown in spring - March until April.

Preprocessing of C & E Data

Since 1946 the C & E data have been recorded in field books during the annual seed multiplication (Note: Since 2011, Personal Digital

Assistants (PDA) are used for data recording in the field making the repeated transfer of these data unnecessary.) Subsequently, these data have been transferred to main card files. However, the data had to be digitized prior to computational analysis. Thus, the data had been manually transferred twice, namely from field books to main card files and then to the computer, increasing the potential of introducing errors.

Hence we checked for spurious values in the recorded traits, and tested the temporal order of the recorded dates. The data was split into winter and spring types based on their classification of growth habit, commonly referred to as 'annuality' among genebanks, in IPK's Genebank Information System (GBIS, http://gbis.ipk-gatersleben.de).

For the identification of potential errors, we subsequently performed an extreme outlier detection using three times the interquartile range. These outliers were checked against the field books and corrected wherever possible. To ensure that any accession provided at most one record per year, we checked for accessions with more than one record in any year. In case of an accession that has been recorded more than once in a year, the records were replaced by a single record with the median of the observations for the subsequent analysis.

For further analysis, the values for plant height (recorded at growth stage Z70 in cm[30],) and thousand grain weight (in g) were used directly, while we computed time periods in days for the flowering time. Thus flowering time is recorded as day of year, i.e., the difference between January 1st and date of flowering (Z65[30],). In Tab. S6, we present the amount of data records for barley and wheat classified according to their annuality. After outlier detection, each collection comprised several thousand accessions and up to five times more records.

Normalized Rank Product on C & E Data

Due to the heterogeneity of the data, descriptive statistics cannot easily be utilized for the data analysis. Instead the recorded values of each trait in each year were ranked to overcome the strong effects of annual differences, for instance inter-annual climate variability, climate change or agricultural practice. Subsequently, we introduced the normalized rank product (NRP) that is based on the idea of the popular rank product analysis, originally proposed for selecting differentially expressed genes from microarray experiment data[31].

Here, we extend the rank product to maintain the relation to the geometric mean of the normalized ranks. The NRP of an accession *a* for a specific trait denoted by NRP_a is defined as,

$$NRP_a = \sqrt[|Y_a|]{\prod_{y \in Y_a} \frac{r_{y,a}}{n_y}},$$

for a specific trait where Y_a is the set of years in which accession *a* has been cultivated and its trait has been recorded, $r_{y,a}$ is the rank of the trait value of accession *a* in year *y*, and n_y is the number of accessions with this trait recorded in year *y*. The NRP ranges between zero (exclusive) and one (inclusive) allowing for comparison between accessions, independent of their number of years of cultivation. The NRPs were computed for the three important traits: flowering time, plant height, and thousand grain weight.

In order to avoid overestimating the influence of accessions that have been infrequently cultivated and exhibiting interesting NRPs by chance, only accessions that have been cultivated and recorded in at least three different years were included in the present study. With this set of assumptions, we were able to compare large parts of the above mentioned collections comprising thousands of accessions.

As an illustrative example, a particular wheat accession has been cultivated in 1946, 1967 and 2009. This plant sample was the first to flower in 1946 and 2009 (rank 1), and the third in 1967 (rank 3). These ranks are divided by the number of all accessions of that species cultivated in each year. The geometric mean is taken over 1946, 1967 and 2009 yielding a NRP for flowering time of this accession ranging between 0 (early flowering) and 1 (late flowering).

Experimental Validation under Field Conditions for Winter Wheat

Based on the NRP and multi-trait optimization (MTO), we were able to find plant samples with contrasting phenotypes, for example winter wheat with extreme flowering time and plant height (Fig. 3A). To validate the accuracy of this procedure, 60 winter wheat accessions (four groups with 15 accessions each) were selected and planted in a field experiment at Gatersleben (51° 49′N, 11° 16′E) during the growing season 2010/2011. We used plots of size 1.5 m × 2.5 m separated by equally sized plots of winter barley and followed local agricultural practice. In the field experiment, we performed three replicates arranged in three blocks. Each block consisted of one plot per selected accession in random order.

Each plot was planted at a density of 80 seeds per m^2. Subsequently, 10 individuals per plot – in total 1,800 individuals – were randomly selected and genetically purified as genebank accessions are not expected to be homogenous and must be purified for genomic studies. Thus all 1800 individuals were covered with bags to prevent cross pollination and phenotyped. The development of single seed descends (SSDs) is a procedure widely used in plant breeding[32]. All individuals were phenotyped for flowering time (Z60[30]) as well as for plant height (Z70[30]), and were finally harvested. Leaf samples for DNA isolation were taken from every plant at Z30[30].

Based on known geographical origin and pedigree information, a subset of 28 accessions (out of 60 accessions mentioned above) was then subselected (6–8 accessions per contrasting group) for an independent second field experiment in 2011/2012 – in order to validate the data from the experiment in 2010/2011. As in the year before, we used plots of size 1.5 m × 2.5 m separated by equally sized plots of winter barley and followed local agricultural practice. Three replicates were performed, arranged in three blocks. Each block consisted of one plot per selected accession in random order. Each plot was planted at a density of 80 seeds per m^2. Full plots were phenotyped for flowering time (Z60[30]) as well as for plant height (Z70[30], Fig. S7).

Extraction of Genomic DNA

Genomic DNA was isolated from silica-dried single leaves of each line with the Qiagen DNeasy Plant Mini Kit (Qiagen, Hilden, Germany), according to the manufacturer's instructions.

Genotyping Major Vernalization Genes for Wheat

Wheat can be divided into spring and winter habit varieties along with a group of intermediate lines known as facultative varieties.

The vernalization-sensitivity (winter habit) alleles, *vrn-A1*, *vrn-B1*and *vrn-D1* are located on chromosomes 5A, 5B and 5D, respectively[33, 34]. Another vernalization gene, *Vrn-3*, located on chromosome 7B, was shown to encode a homolog of *Arabidopsis FT*, and was named *VRN3*[35].

Aiming at reducing any bias for flowering time analysis based on growth habit, in total, 60 purified individuals (1 to 3 plants per accession) were randomly selected among the 28 contrasting accessions subselected for the second field experiment (see above) and genotyped at four major vernalization loci (*Vrn-A1*, *Vrn-B1*, *Vrn-D1* and *Vrn-B3*) using allele-specific molecular assays. Following the experimental protocols[22, 23, 24, 35, 36, 37, 38, 39, 40] (Tab. S3), all accessions were confirmed as true winter wheat harboring only vernalization-sensitive alleles providing strong support for the flowering time analysis.

Resequencing the Major Photoperiod Response Locus *Ppd-D1*

Sixty contrasting accessions (represented by 96 plants, 1 to 3 plants per accession) were resequenced at *Ppd-D1*. In order to develop wheat D genome specific primer combinations, sequence information was obtained from the International Wheat Genome Sequencing Consortium (IWGSC, http://www.wheatgenome.org/) and NCBI GenBank (DQ885766.1).

The Primer3 online software (Primer3 v. 0.4.0 (http://frodo.wi.mit.edu/primer3/) was used to design primers. Oligonucleotides were purchased from Eurofins MWG Operon, Ebersberg, Germany. One c. 5920 bp genomic region completely covering *Ppd-D1* was amplified by locus-specific PCR primers from genomic DNAs (Tab. S4). 5′-UTR, exons, introns and 3′-UTR were analyzed, as start and stop codon of the reference (DQ885766.1) were located at alignment position 2150 and 5299, respectively. The exons were located at 2150 to 2321, 2439 to 2603, 2712 to 2845, 3041 to 3196, 3648 to 3825, 3932 to 4349, 4432 to 5089, 5200 to 5301. Specificity and chromosomal localization of PCR products were confirmed by Nulli-tetrasomic (NT) lines (N2A-T2B, N2B-T2D and N2D-T2B)[41] (Fig. S8).

PCR amplification was performed in 20 μl reaction volume. Templates were purified and sequenced directly on both strands on an Applied Biosystems (Weiterstadt, Germany) ABI Prism 3730 xL sequencer using BigDye terminators as described in[42]. Their DNA sequences were determined using primers designed for amplification and internal primers (Tab. S4).

Single Nucleotide Polymorphism (SNP)-Detection

DNA sequences were processed with AB DNA Sequencing Analysis Software 5.2 and later manually edited by Sequencher software v5.0 (Gene Codes Corp.). Sequence alignments were generated with MAFFT webserver (http://www.ebi.ac.uk/Tools/msa/mafft/) using default parameters except "perform ffts" which was set to "genafpair". Subsequently, the multiple sequence alignment was manually modified. Filtering sequences with low quality regions, 67 sequences were contained in the final alignment comprising c. 5920 bp. The heterozygous state at the deletion in the promoter region was indicated by inserting poly-N. In close analogy, the heterozygote state for the transposable element in the first intron was indicated by inserting poly-N.

Allelic haplotypes were defined by DNASP 5.10.01 as described in[42] (Tab. S5). All identified singletons were confirmed afterwards by additional two independent amplifications and sequencing.

ACKNOWLEDGEMENTS

We thank M. Grau, M. Mildner, M. Oppermann, G. Schütze, and R. Selbig for technical assistance with the C & E data. We thank P. Abraham, S. Dreiβig, B. Dubsky, H. Giraud, H. Harms, C. Kehler, U. Krajewsky, K. Neumann, M. Nix, P. Schreiber, and K. Wolf for excellent assistance in the field experiments. We thank E. Andeden and K. Niedung for DNA analysis and S. Kilian for kindly preparing Figure 1. We are grateful to T. Altmann, B. Bauer, G. Coupland, M. Friedel, W. Gruissem, H. Riegler, M. Röder, R. Schnee, M. Seifert, R. Sharma, S. Singh, N. Stein, M. Strickert and N. von Wirén for discussions. The authors would like to acknowledge the support given by the Ministry of Culture of Saxony-Anhalt (grant XP3624HP/0606T) to JK and SF, and the German Science Foundation Priority Programme SPP1530 to BK.

REFERENCES

1. Takeda, S. & Matsuoka, M. Genetic approaches to crop improvement: responding to environmental and population changes. *Nat Rev Genet* 9, 444–457 (2008).
2. Foley, J. A. *et al.* Solutions for a cultivated planet. *Nature* 478, 337–42 (2011).
3. Esquinas-Alcazar, J. Protecting crop genetic diversity for food security: political, ethical and technical challenges. *Nat Rev Genet* 6, 946–953 (2005).
4. Fitter, A. H. & Fitter, R. S. R. Rapid changes in flowering time in British plants. *Science* 296,1689–1691 (2002).
5. Godfray, H. C. J. *et al.* Food Security: The challenge of feeding 9 billion people. *Science*327, 812–818 (2010).
6. Iglesias, A., Quiroga, S. & Diz, A. Looking into the future of agriculture in a changing climate.*Europ Rev Agricult Econ* 38, 427–447 (2011).
7. Tester, M. & Langridge, P. Breeding Technologies to Increase Crop Production in a Changing World. *Science* 327, 818–822 (2010).
8. Graybosch, R. A. & Peterson, C. J. Genetic Improvement in Winter Wheat Yields in the Great Plains of North America, 1959–2008. *Crop Sci* 50, 1882–1890 (2010).
9. Brisson, N. *et al.* Why are wheat yields stagnating in Europe? A comprehensive data analysis for France. *Field Crops Res* 119, 201–212 (2010).
10. Hoisington, D. *et al.* Plant genetic resources: What can they contribute toward increased crop productivity? *Proc Nat Acad Sci* 96, 5937–5943 (1999).
11. Kilian, B. & Graner, A. NGS technologies for analyzing germplasm diversity in genebanks.*Brief Funct Genom* 11, 38–50 (2012).
12. Snook, L. K., Dulloo, M. E., Jarvis, A., Scheldeman, X. & Kneller, M. [Crop germplasm diversity: The role of gene bank collections in facilitating adaptation to climate change]. *Crop Adaptation to Climate Change* [Yadav S. S., Redden R., Hatfield J. L., Lotze-Campen H., & Hall A., (eds.)] [495–506] (Wiley-Blackwell, 2011).

13. Tanksley, S. D. & McCouch, S. R. Seed banks and molecular maps: Unlocking genetic potential from the wild. *Science* 277, 1063–1066 (1997).
14. Amano, T., Smithers, R. J., Sparks, T. H. & Sutherland, W. J. A 250-year index of first flowering dates and its response to temperature changes. *Proc Royal Soc B-Biol Sci* 277,2451–2457 (2010).
15. Cleland, E. E., Chuine, I., Menzel, A., Mooney, H. A. & Schwartz, M. D. Shifting plant phenology in response to global change. *Trends Ecol Evol* 22, 357–365 (2007).
16. Craufurd, P. Q. & Wheeler, T. R. Climate change and the flowering time of annual crops. *J Exp Bot* 60, 2529–2539 (2009).
17. Estrella, N., Sparks, T. H. & Menzel, A. Trends and temperature response in the phenology of crops in Germany. *Global Change Biol* 13, 1737–1747 (2007).
18. Hu, Q., Weiss, A., Feng, S. & Baenziger, P. S. Earlier winter wheat heading dates and warmer spring in the US Great Plains. *Agri Forest Meteorol* 135, 284–290 (2005).
19. Jia, J. *et al. Aegilops tauschii* draft genome sequence reveals a gene repertoire for wheat adaptation. *Nature* 496, 91–95 (2013).
20. Brenchley, R. *et al.* Analysis of the bread wheat genome using whole-genome shotgun sequencing. *Nature* 491, 705–710 (2012).
21. Ling, H.-Q. *et al.* Draft genome of the wheat A-genome progenitor *Triticum urartu*. *Nature*496, 87-90 (2013).
22. Beales, J., Turner, A., Griffiths, S., Snape, J. W. & Laurie, D. A. A pseudo-response regulator is misexpressed in the photoperiod insensitive Ppd-D1a mutant of wheat (*Triticum aestivum* L.). *Theor Appl Genet* 115, 721–733 (2007).
23. Wilhelm, E. P., Turner, A. S. & Laurie, D. A. Photoperiod insensitive *Ppd-A1a* mutations in tetraploid wheat (*Triticum durum* Desf.). *Theor Appl Genet* 118, 285–294 (2009).
24. Nishida, H. *et al.* Structural variation in the 5′ upstream region of photoperiod-insensitive alleles *Ppd-A1a* and *Ppd-B1a* identified in hexaploid wheat *(Triticum aestivum* L.), and their effect on heading time. *Mol Breed* 1–11 (2012).
25. Shaw, L. M., Turner, A. S., Herry, L., Griffiths, S. & Laurie, D. A. Mutant alleles of*Photoperiod-1* in wheat (*Triticum aestivum* L.)

that confer a late flowering phenotype in long days. *PLoS One* 8, e79459 (2013).

26. Brown, A. H. D. Core collections: a practical approach to genetic resources management.*Genome* 31, 818–824 (1989).
27. Winfield, M. O. *et al.* Targeted re-sequencing of the allohexaploid wheat exome. *Pl Biotech J*10, 733–742 (2012).
28. Knüpffer, H. [Triticeae genetic resources in *ex situ* genebank collections]. *Genetics and Genomics of the Triticeae* [Feuillet C., & Muehlbauer G. J. (eds.)] [31–79] (Springer, 2009).
29. Börner, A. Preservation of plant genetic resources in the biotechnology era. *Biotechnol J* 1,1393–1404 (2006).
30. Zadoks, J. C., Chang, T. T. & Konzak, C. F. Decimal Code for Growth Stages of Cereals.*Weed Res* 14, 415–421 (1974).
31. Breitling, R., Armengaud, P., Amtmann, A. & Herzyk, P. Rank products: a simple, yet powerful, new method to detect differentially regulated genes in replicated microarray experiments. *Febs Lett* 573, 83–92 (2004).
32. Acquaah, G. *Principles of Plant Genetics and Breeding,* (Wiley, 2012).
33. Worland, T. & Snape, J. [Genetic basis of worldwide wheat varietal improvement]. *The World Wheat Book: A history of wheat breeding.* [Angus W. J., Bonjean A. P., & Van Ginkel M. (eds.)] [61–67] (Lavoisier Publishing, Paris, 2001).
34. Yan, L. [The Flowering Pathway in Wheat]. *Wheat Science and Trade* [Carver B. F. (ed.)] [57–72] (Wiley-Blackwell, 2009).
35. Yan, L. *et al.* The wheat and barley vernalization gene *VRN3* is an orthologue of FT. *Proc Nat Acad Sci* 103, 19581–19586 (2006).
36. Bentley, A. R. *et al.* Frequency of photoperiod-insensitive *Ppd-A1a* alleles in tetraploid, hexaploid and synthetic hexaploid wheat germplasm. *Pl Breed* 130, 10–15 (2011).
37. Díaz, A., Zikhali, M., Turner, A. S., Isaac, P. & Laurie, D. A. Copy number variation affecting the *Photoperiod-B1* and *Vernalization-A1* genes is associated with altered flowering time in wheat (*Triticum aestivum*). *PLoS ONE* 7, e33234 (2012).
38. Guo, Z. A., Song, Y. X., Zhou, R. H., Ren, Z. L. & Jia, J. Z. Discovery, evaluation and distribution of haplotypes of the wheat *Ppd-D1* gene. *New Phytol* 185, 841–851 (2010).

39. Yan, L. *et al.* Allelic variation at the *VRN-1* promoter region in polyploid wheat. *Theor Appl Genet* 109, 1677–1686 (2004).
40. within the first intron in *VRN-1* are associated with spring growth habit in barley and wheat. *Mol Genet Genom* 273, 54–65 (2005).
41. Sears, E. R. [Nullisomic-tetrasomic combinations in hexaploid wheat]. *Chromosome Manipulations and Plant Genetics* [Riley R., & Lewis K. R. (eds.)] [29–45] (Springer Science+Business Media, LLC, 1966).
42. Rozas, J. DNA sequence polymorphism analysis using DnaSP. *Meth Mol Biol* 537, 337–350(2009).

Chapter 2

A Phylogenetic Analysis of the Grape Genus (Vitis L.) Reveals Broad Reticulation and Concurrent Diversification during Neogene and Quaternary Climate Change

Mirza Hasanuzzaman[1], Kamrun Nahar[2,3], Md. Mahabub Alam[2], Prasanta C. Bhowmik[4], Md. Amzad Hossain[5], Motior M. Rahman[6], Majeti Narasimha Vara Prasad[7], Munir Ozturk[8], and Masayuki Fujita[2]

Yizhen Wan[1], Heidi R Schwaninger[2], Angela M Baldo[2,3], Joanne A Labate[2], Gan-Yuan Zhong[2,3], and Charles J Simon[2]

[1]College of Horticulture, Northwest A&F University, Yangling, Shaanxi 712100, People's Republic of China

[2]US Department of Agriculture, Agriculture Research Service, Plant Genetic Resources Unit, New York State Agricultural Experiment

Station, Cornell University, Geneva, NY 14456, USA

[3]US Department of Agriculture, Agriculture Research Service, Grape Genetic Research Unit, New York State Agricultural Experiment Station, Cornell University, Geneva, NY 14456, USA

ABSTRACT

Background

Grapes are one of the most economically important fruit crops. There are about 60 species in the genus *Vitis*. The phylogenetic relationships among these species are of keen interest for the conservation and use of this germ plasm. We selected 309 accessions from 48 *Vitis* species, varieties, and outgroups, examined ~11 kb (~3.4 Mb total) of aligned nuclear DNA sequences from 27 unlinked genes in a phylogenetic context, and estimated divergence times based on fossil calibrations.

Results

Vitis formed a strongly supported clade. There was substantial support for species and less for the higher-level groupings (series). As estimated from extant taxa, the crown age of *Vitis* was 28 Ma and the divergence of subgenera (*Vitis* and *Muscadinia)* occurred at ~18 Ma. Higher clades in subgenus*Vitis* diverged 16 – 5 Ma with overlapping confidence intervals, and ongoing divergence formed extant species at 12 – 1.3 Ma. Several species had species-specific SNPs. NeighborNet analysis showed extensive reticulation at the core of subgenus *Vitis* representing the deeper nodes, with extensive reticulation radiating outward. Fitch Parsimony identified North America as the origin of the most recent common ancestor of extant *Vitis* species.

Conclusions

Phylogenetic patterns suggested origination of the genus in North America, fragmentation of an ancestral range during the Miocene, formation of extant species in the late Miocene-Pleistocene, and differentiation of species in the context of Pliocene-Quaternary tectonic and climatic change. Nuclear SNPs effectively resolved relationships at and below the species level in grapes and rectified several misclassifications of accessions in the repositories. Our results challenge current higher-level classifications, reveal the abundance of genetic diversity in the genus that is potentially available for crop improvement, and provide a valuable resource for species delineation, germplasm conservation and use.

BACKGROUND

Grapes (*Vitis* spp.) are one of the world's most economically valuable fruit crops [1]. They are widely used for wine, table grapes, raisins, juice, and spirits; recent trends have also focused on antioxidants and healthful products derived from grapes. V*itis vinifera* L. subsp. *vinifera* (referred to as *V. vinifera* here after) is the most widely cultivated grape species but its productivity was historically limited due to its susceptibility to pests, diseases, and abiotic stress such as cold [2]. Genes from wild grape germplasm have been used to improve biotic and abiotic tolerance and resistance in cultivated grapes.

Centers of grapevine diversity are found in the southeastern US [3-5] and East Asia [4,6]. Up to 30 species are native to a vast area in eastern Asia, China, Japan and Java, two species across middle Asia and Europe, and up to 28 species across the eastern and southwestern US and Mexico [2] (Figure 1). expands discussion of the biogeographic background. The genus *Vitis* is divided into two subgenera: *Muscadinia* Planch. ($2n=40$, one or two species) and *Vitis* Planch. ($2n=38$, the remaining species). Additional divisions within *Vitis* are "series" which are subgeneric groupings that have been used historically in the systematics of *Vitis*. They rank below "sections" that are more commonly used in plant systematics for groups of species. Although all wild species are considered diploid, there is evidence of hexaploidization in their distant past [7-9] that is shared with all rosids [9]. The two subgenera are nearly reproductively isolated while the species within subgenus *Vitis* are interfertile. All species are dioecious except *V. vinifera* which has hermaphroditic flowers, and *V. rotundifolia* Michx. which segregates for this trait. Many species have overlapping distributions, thus natural hybridization would occur were it not for ecological and phenological barriers [3,10,11]. Not surprisingly, the classification of *Vitis* is confused in part due to the lack of agreement among systematic botanists as to what constitutes a true species and because of extreme morphological variation within the species [2,3,7]. This has led to many extraneous species names [7,12]. The systematics of *Vitis* is based primarily on morphology [13] and molecular methods have only recently been used to study this taxonomic problem.

Figure 1: Native geographic distribution of the genus *Vitis* (grey shading1) and geographic regions of origin of *Vitis* species used in this study. Dashed lines indicate southern borders of the polar ice cap during the most recent ice age[2]. Dash-dot lines indicate ice age refugia of the forest flora[2]. Areas labeled 1 through 4 were used in ancestral area optimization (reversible parsimony). Redrawn from [1]Alleweldt et al. [7],[2]Reinig [14].

Most previous molecular studies on the evolutionary history of *Vitis* were limited in taxonomic scope or marker choice [15-32]. Studies most similar in goals and pertinent to the present study were conducted by Aradhya et al. [33], Nie et al. [34], Liu et al. [35], and Zecca et al. [36]. Aradhya et al.[33] obtained a taxon sample similar to the present study and used SSR and AFLP markers to study genetic diversity within

Vitis. These markers have limited value for phylogeny reconstruction [37] and dating divergences was not attempted. Nie et al. [34] and Liu et al. [35] provided well-reasoned paleontological dates to estimate divergences in Vitaceae. These calibration points were applied in the present study. Zecca et al.'s [36] chronogram is a tantalizing expansion of the *Vitis* component of Nie et al.'s [34] chronogram. Their inferences were limited by the small number of markers and the limited variability available in those markers, which did not fully resolve the tip clades. Further, limited intra-specific replication (sampling) limited the ability to make species-level inferences.

Adding more data can be useful for resolving difficult phylogenies that were based on a few genes [38]. The present study attempted to improve on three aspects of previous phylogeographical studies of *Vitis* by more extensive sampling of the nuclear genome, the species, and intraspecific variation. This study developed and used 27 nuclear gene markers and sequenced 309 accessions of 48 *Vitis* species, varieties, and four out-groups to: 1) reconstruct a phylogenetic hypothesis of the genus *Vitis*, 2) date important time points in the evolution of *Vitis*, 3) elucidate the biogeographic history of the genus, and 4) evaluate systematics of *Vitis* within the framework of phylogeny.

RESULTS

Molecular Characteristics of the Nuclear Sequences of *Vitis*

Most *Vitis* accessions had complete sequence or had minimal missing data. Indel sequences in the 27 gene markers were unambiguous and easy to align. The starting alignment matrix for the 27 gene markers and all 309 accessions was 11,437 bp long. Gap coding for Maximum Parsimony added 304 characters. Amino acid coding sequence accounted for 5,690 nt, 3′ or 5′ untranslated regions for 4,074 nt, and introns for 1,036 nt. Because Trees by New Technology (TNT) does not output the number of parsimony-informative characters, we report the number of unique site patterns from Bayesian Evolutionary Analysis by Sampling Trees (BEAST) and Randomized Accelerated Maximum Likelihood (RAxML). The 52-OTU matrix had 1,855 unique site patterns; the 273-OTU matrix had 2,510 unique site patterns. Under the uncorrelated log-normal relaxed molecular clock, estimated from three combined runs in BEAST and calibrated with three fossil dates, the mean rate of substitution (meanRate) in the data set was 8.249×10^{-4} per million years (Effective Sample Size (ESS)=959) and a coefficient of variation (CV) of 0.896 (ESS=1284). Dividing this rate by the size of the data set (11,437 nt), the average rate of substitution in this data set was 7.2×10^{-8} substitutions per site per million years.

Extent of Reticulation and Network in *Vitis*

Ancestral polymorphism with subsequent lineage sorting is difficult to distinguish from reticulation based on a phylogenetic pattern [38-42]. In our study, six (22%) gene fragments showed significant tests for recombination based on the Phi Test [43]: fragments 1313 (P = 0.033), 1973 (P = 0.033), 2415 (P = 0.006), 5069 (P = 0.033), 7022 (P = 0.047), 7029 (P = 0.019). When these fragments were excluded, the concatenated matrix still showed recombination, as expected (P = 0.000), representing independent assortment of markers. Lanier and Knowles [44] found that, in species-tree estimation, the gain of accuracy from sampling additional loci and/or individuals always exceeded inaccuracies related to recombination. Thus, in the present work no genes were excluded from subsequent analyses based on evidence of recombination.

We used networks to better visualize the conflicts represented by the high levels of homoplasy. The consensus network [45] of the 26 individual gene trees indicated that few splits were common to multiple trees. The NeighborNet [45] of the concatenated 273-OTU matrix (Figure 2) showed extensive conflict at the core of subg. *Vitis* representing the deeper nodes, and extensive conflict radiating outward.

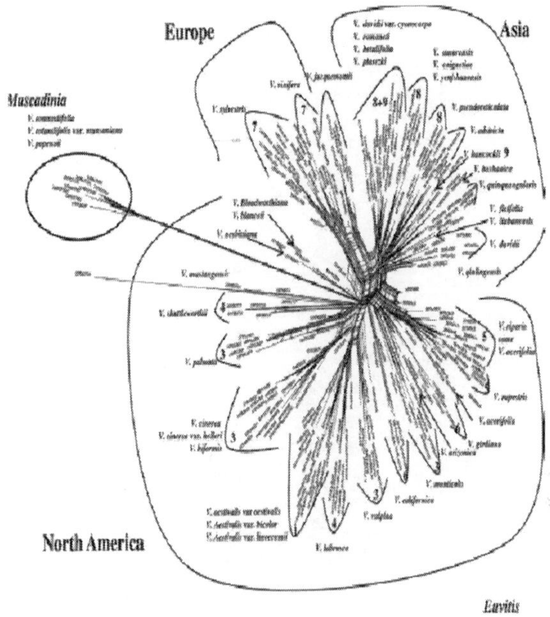

Figure 2: The NeighborNet of 273 accessions based on 27 concatenated nuclear gene fragments. Numbers indicate the series to which species have been recognized 1: *Aestivales* (Planchon); 2:*Cinerascentes* (Planchon); 3: *Cordifoliae* (Munson); 4: *Labruscae* (Planchon); 5: *Ripariae* (Munson); 6: *Occidentales* (Munson); 7: *Viniferae* (Planchon); 8:*Flexuosae* (Galet); 9: *Spinosae* (Galet).

Estimation of Divergence Times in *Vitis*

With the stem age of Vitaceae constrained at 90.7 ± 1.0 Ma, the stem age of *Vitis* at 58.5 ± 5 Ma, and the divergence of *V. labrusca* and its closely related North American relatives at 5.75 ± 0.5 Ma, the crown age of *Vitis* was estimated at 28.32 Ma (95% Highest Posterior Density (HPD) 41.25, 16.23), the crown age of subg. *Vitis* at 17.82 Ma (95% HPD 26.71,10.14), and the stem ages of most species fell between 11 and 1.3 Ma. The ten individual runs continued for 61.31 million to 89.21 million steps, and estimated the crown age of subg. *Vitis* between 17.285 Ma and 18.238 Ma. The Bayesian divergence times in Figure 3 were estimated from three combined, unpartitioned runs using BEAST. The maximum clade credibility tree with mean estimates of divergence time for all nodes and associated posterior probabilities were also obtained.

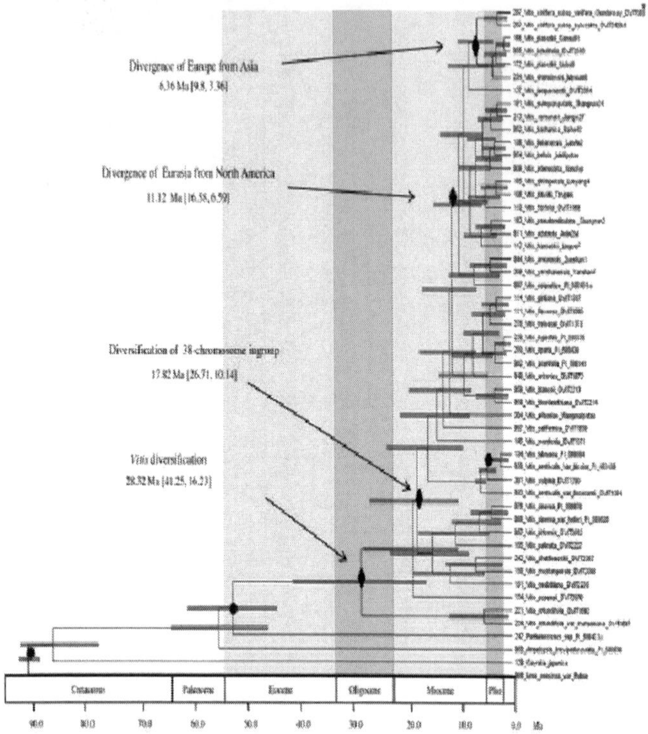

Figure 3: Chronogram of Bayesian divergence time estimates of *Vitis*diversification based on 27 concatenated nuclear gene fragments inferred using the BEAST software. Grey bars represent the 95% Highest Posterior Density (HPD) intervals of nodal age in million years. Calibration points are indicated with filled circles. Significant evolutionary events are indicated with black diamonds. Asterisk indicates inclusion of a clonally propagated cultivar that may affect the local divergence estimate. 4 and 5 show nodal ages and posterior probabilities for all nodes in this tree.

Phylogenetic Analyses in *Vitis*

Maximum Likelihood

ML analyses of 26 single genes analyzed independently without missing data yielded 26 very poorly resolved trees. No significant conflict was observed under ML, thus the sequences were concatenated.

The 20 identical (except for the starting seed) partitioned rapid ML runs using the 273-OTU matrix produced a range of maximum likelihood values of -42511to -42380. The final search on the same matrix following the bootstrap search yielded the highest likelihood value: -42358. This tree was used in comparisons with results from other methods using a cartoon, i.e., a simplified version with collapsed terminal clades. Bootstrap values are reported for selected nodes (Figures 4 and 5) and all nodes.

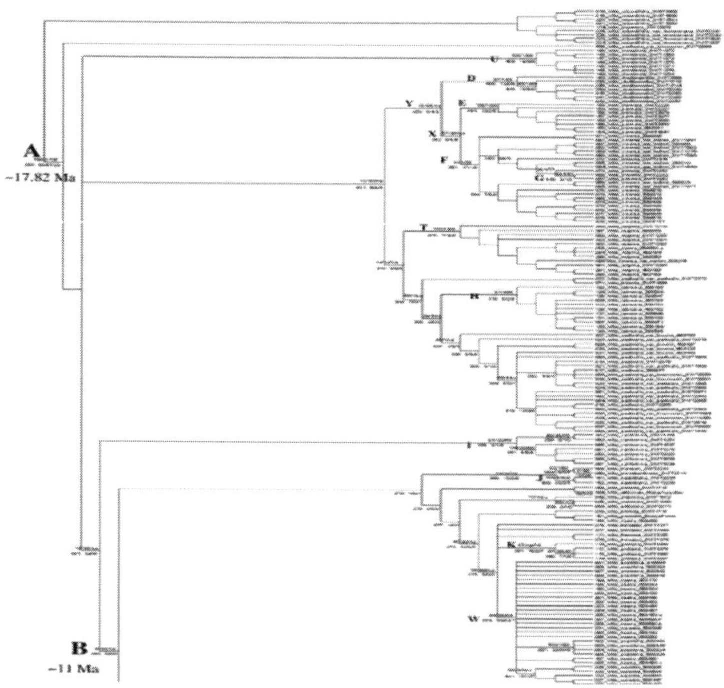

Figure 4: Hypothesis of phylogenetic relationships among *Vitis*species. North America. Strict consensus of 282 most parsimonious trees from four driven searches with support values for selected nodes. Above branches: (Maximum Likelihood BS/ Bayesian PP/ Maximum Parsimony BS). Values can range from 1-100 in ML, 0-1 in BA, 50-100 in MP. Below branches: (node number: branch length/No unique characters/No genes in support of node). The annotation "319: 8/2/6" means that node number 319 has 8 character changes of which 2 are unique to that node, and 6 genes contributed changes. "na" indicates absence of a value in the specific support category. Branch lengths, bootstrap supports. Nodes labeled A-Y are discussed in the text. Figure 4 continues in Figure 5.

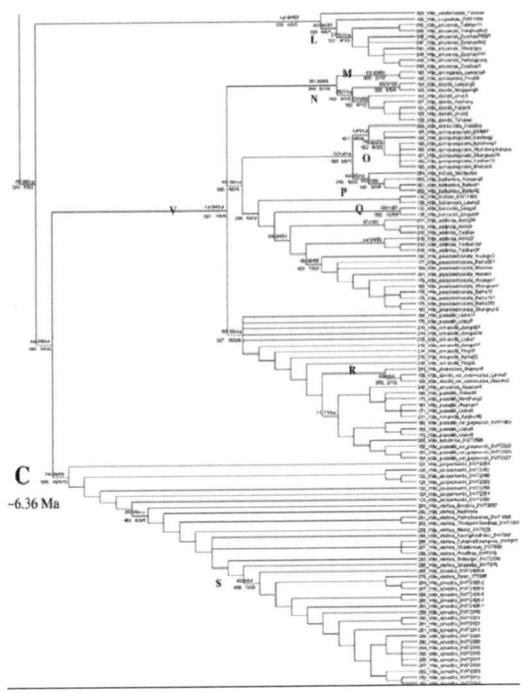

Figure 5: Hypothesis of phylogenetic relationships among *Vitis*species. Eurasia.Continuation of Figure 4.

Bayesian

The two runs timed out at 48 mil and 50 mil generations with good effective sample size (ESS; >>200), but they did not converge on the exact same phylogenetic hypothesis. Both trees were highly concordant for the species-level clades but differed on the specific placement of some clades. Because the 48 mil run had a higher mean log likelihood of the cold chain (LnL) after burnin, the posterior probabilities from this run were used to summarize supports (Figures 4 and 5). Overall comparisons of relationships above the species level were facilitated using the cartoon of this tree.

Maximum Parsimony

TNT's driven search function produced most parsimonious (MP) trees of the same length in all four searches. The individual MP trees had a score of 4398, consistency index (CI) of 0.360, and retention index (RI) of 0.790. The strict consensus of all 282 MP trees from all four searches had a score of 4580, CI=0.346, and RI=0.776. The strict consensus tree was simplified by collapsing terminal clades (Figure 6) for comparisons. The full tree was annotated with support values on selected nodes of interest (Figures 4 and 5, see figure legend). All node numbers are shown in. The MP bootstrap tree with supports >50% was illustrated

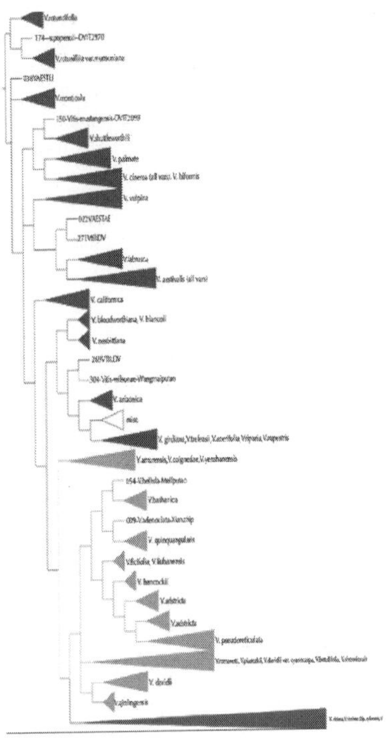

Figure 6: Simplified version (cartoon) of the MP strict consensus tree. Blue = North and Central American accessions, Green = Asian accessions, Red = European accessions. Represent cartoons of the ML and BA trees, respectively.

Ancestral Area Analysis

Fitch parsimony identified Eastern/Southeastern North America (that also included Mexico and the Caribbean) as the origin of the most recent common ancestor of *Vitis* based on the strict consensus tree

Systematics of Vitis spp.

It was evident that the patterns were similar among the cartoons of the strict MP consensus (Figure 6), the highest ML value cladogram and the Bayesian cladogram. The clades of Eurasian species were nested in North America as a monophyletic clade. The series *Precoces*Munson (containing *V. riparia* Michaux, *V. acerifolia* Raf., *V. rupestris* Scheele) together with *V. arizonica*Engelm. (belonging to the series *Occidentales* Munson, *V. blancoii* Munson, *V. bloodworthiana* Comeaux, *V. Xtreleasii* Munson ex L. H. Bailey and *V. girdiana* Munson formed the sister clade. *V. labrusca* and *V. aestivalis* Michx. were grouped together as were *V. cinerea* (Engelm.) Engelm. ex Millardet, *V. palmata* Vahl, *V. shuttleworthii* House and *V. mustangensis* Buckley. All three analyses did not group *V. monticola* Buckley or *V. californica* Benth. with other species.

The position of several clades differed among the searches. This was a notable characteristic of this data set. Among the different BA, ML and MP trees shown, the OTU composition of species clades in general was quite consistent and well supported, but a few clade or species positions were inconsistent. Correspondingly, clades above the species level were often poorly supported. For example, *V. yenshanensis* J. X. Chen was grouped with *V. amurensis* Rupr. and *V. coignetiae* Pulliat ex Planch. in MP and BA analysis and in the network but not in ML. Similarly, the *amurensis/coignetiae*clade was basal to Eurasia in MP and BA, but placed within the Asian clade in ML and in the network. *V. nesbittiana* Comeaux were grouped with *V. mustangensis* and *V. shuttleworthii* in BA and ML but grouped with *V. bloodworthiana* and *V. blancoii* in MP and in the network.

To better understand clade support, we further investigated the synapomorphies defining clades of interest. The characters supporting nodes of interest and their level of homoplasy and gene source were summarized based on MP. Summary statistics for specific nodes investigated are shown below the branches in Figures 4 and 5. Many species showed good support with high bootstrap values and posterior probabilities, presence of node-specific characters and support from multiple genes (letters in parentheses refer to marked nodes in Figures 4 and 5): *V. shuttleworthii* (D),*V. palmata* (E), *V.cinerea* (all varieties and including *V. biformis* Rose) (F), *V. biformis* (G), *V. labrusca*(H),

V. californica (I), *V. nesbittiana* (J), *V. girdiana* (K), *V. amurensis* (L), *V. qinlingensis* (M), *V. davidii*(Rom. Caill.) Foëx (N), *V. quinquangularis* Rehd. (O), *V. bashanica* P.C. He (P), *V. hancockii* Hance (Q), *V. davidii var. cyanocarpa* (Rom. Caill.) Foëx (R; and not grouping with *V. davidii*), *V. vinifera* ssp. *sylvestris* (C. C. Gmel.) Hegi. (Referred to as *V. sylvestris* hereafter) (S). the species *V. vulpina* L. (T) and *V. monticola* (U) each formed reliable species-level clades even in the absence of a species-specific SNP. Many higher level relationships were supported by a few characters and were of poor quality leading to the labile topology among major clades. Examples were the nodes defining the split between North America and Asia (B), Europe (with *V. jacquemontii* R. Parker) and China (C), and China without *V. amurensis* (V). These nodes had five to seven supporting characters that were frequently highly homoplasious (Figure 4;. The notable exception was the branch separating the *Muscadinia* from subg. *Vitis* (Figure 4 node A). It was supported by 93 characters of which 57 (61%) showed no homoplasy and represented 20 of the 27 (74%) gene markers. Other supported higher clades were the *V. cinerea-V. palmata* clade (X in Figure 4) and the (*V. mustangensis, V. shuttleworthii, V. palmata, V. cinerea* with *V. biformis*) clade (Y). Both nodes had a node-specific SNP, although bootstrap support and posterior probabilities were more consistent with other poorly supported higher-level clades.

The markers were informative in characterizing intra-specific variation in some species. The MP branch lengths, MP and ML bootstrap supports and BA posterior probabilities supported intraspecific groupings well with non-zero branch lengths present in many species such as *V. shuttleworthii, V. monticola, V. californica, V. palmata, V. labrusca, V. cinerea, V. aestivalis, V. sylvestris, V. adstricta,* and *V. davidii*. Due to space constraints these supports were not summarized in (Figures 4 and 5).

DISCUSSION

Problems of Phylogenetic Study in *Vitis*

Based on MP, ML, and BA phylogenetic reconstruction methods, the nuclear DNA dataset in this study had extensive variation to address genus-wide relationships in *Vitis*. The markers characterized intraspecific variation, defined most species, and strongly supported subg. *Vitis*. However, many of the relationships and deeper nodes (above species within subg. *Vitis*) were characterized by low bootstrap values and were often supported by few characters with high homoplasy. Low clade support and high homoplasy may be caused by insufficient data, parallel changes, reversals and convergences, as well as different histories of genes caused by lineage sorting and reticulation [40,46], and different clade sizes[47]. The ascertainment bias (the systematic distortion in measuring the true frequency of SNPs due to sampling) introduced in the marker development phase may have selected markers with insufficient variation outside *V. vinifera* or phylogenetic depth, and differences may be distorted due to domestication. However, we re-sequenced whole fragments, rather than genotyping *a priori* identified SNPs, thus included additional markers not restricted by our selection criteria of intermediate variability (see Methods). Further, the deepest node was very well supported both in number and quality of characters, illustrating that there was sufficient phylogenetic signal available to define subg. *Vitis*. Similarly, several species (e.g. *V. shuttleworthii, V. labrusca, V. palmata, V. hancockii, V. quinquangularis*) were well supported by multiple non-homoplasious characters. Thus, simple lack of data was not the definitive reason for poor support of deep nodes. Additional data, using the Vitis9KSNP array [22] or others in development, markers developed by Lijavetzki et al. [20] and Vezzulli et al. [30] or next generation sequencing (NGS) may resolve this phylogenetic problem. However, these data will certainly add more noise (homoplasious characters). In addition, ascertainment of homology in the data set created by NGS is very intractable because of complex paleopolyploidization and gene duplication in the grapevine genome [9]. It is possible that additional data from non-recombining chloroplast or mitochondrial DNA might add stable characters deeper within the tree. However, the literature [19,23,25,29,36,48] suggests that nucleotide substitution rates in these datasets may be too slow to add much intra-generic information. Species reticulation and incomplete lineage sorting would still present a challenge. Parallel changes, reversals, and convergence are likely minor contributing factors to the observed homoplasy due to the shallow phylogenetic depth of this study involving moderate levels of evolutionary time; these factors were further minimized by locus selection criteria.

Ancestral Polymorphism, Reticulation and Incomplete Lineage Sorting

Homoplasy due to incomplete sorting of ancestral alleles is more likely when the time between lineage splitting is short (short branch especially when deep in the tree [49]) and the effective population size is large [50]. The present estimates of divergence times showed that splitting events between the deeper clades occurred almost simultaneously within subg. *Vitis*. Myles et al. [22] found significant degrees of shared polymorphisms between North American wild grapevine species and European cultivated species, suggesting that grapevine species maintained large effective population sizes since their geographic isolation millions of years ago. Further, the linkage disequilibrium in *V. vinifera* is very low and haplotype blocks are very short [20-22,51], indicating significant historical recombination within the species [22]. There was significant recombination in several genes and in the concatenated dataset. Thus, the conflict in the NeighborNet (Figure 2) can be interpreted as evidence of shared ancestral polymorphisms mixed with reticulation and lineage sorting. The shared ancestral polymorphisms may be the cause of the central knot of conflict (an ancestral ocean of polymorphisms and reticulation) represented by the tight central mass of splits that represent incompatible and ambiguous signals in the data set [45], with the radiating splits representing progressive lineage sorting and reticulation within the lineages. Reticulating events include hybridization, recombination and horizontal gene transfer [45]. The first two were likely major factors in the evolution of *Vitis*, while horizonatal gene transfer was an unlikely mechanism. We conclude that extensive reticulation deep in the tree and incomplete lineage sorting are the likely reason for the lack of support at higher level nodes.

Time Frame of *Vitis* Diversification

Our molecular dating is close to fossil estimates and distributional inferences that place the origination of *Vitis* into the Paleogene. The beginning of diversification among the extant taxa (crown age) in our tree was estimated at 28.32 Ma (95% HPD 41.25 Ma - 16.23 Ma). This was earlier than estimated by Nie et al. [34] who assessed it at about 8 Ma with a large 95% HPD, or Liu et al. [35] who assessed it at about 12 Ma (~22 Ma – 6 Ma), or Zecca et al. [36] with an estimated mean age of 18.60 Ma (28.79 Ma - 9.50 Ma) or 19.05 Ma (29.07 Ma - 10.2 Ma), depending on the details of their analyses. Estimates by all three [34-36] were predominantly based on (non-recombining) chloroplast sequence with 116, 1258, and 54 parsimony informative sites respectively, two studies [34,36] had one nuclear marker with 135 and 41 parsimony informative sites respectively, and each combined data set was <6000 nt, compared to the ~11,440 nt of the present study. Recombination in the present data possibly caused the estimate of the height of the tree to be greater, thus resulting in more ancient age estimates [44]. The use of the distant outgroup *Leea* may have caused problems in dating because of missing data and multiple substitutions. The inclusion of additional more closely related outgroups could improve the accuracy of the inferred dates [52]. The present estimate of 28.32 Ma appears a reasonable age considering that the estimate was associated with a large CI that reached back to 41 Ma and the oldest reliable *Vitis* seed date to the Paleocene (65.5 Ma - 55.8 Ma) [53,54]. The stem of *Vitis* did reach back that far (Figure 3). The diversification of subg. *Vitis* (Node A) was presently estimated at 17.82 Ma (26.71 Ma - 10.14 Ma), a node that was dated at 6.31 or 6.55 Ma (9.5 Ma -3.5 Ma, 9.34 Ma - 4.50 Ma) as node B in Zecca et al. [36], about 5.5 Ma (about 10 Ma - 2.5 Ma) by Nie et al. [34] and about 7 Ma (11 Ma - 5 Ma) by Liu et al. [35]. Higher level intra-North American divergences fell between 15.75 Ma -8.96 Ma. The divergence of Eurasia from North and Central America (the Asian- North American disjunction, node B in Figure 4) was 11.12 Ma (16.58 Ma –6.59 Ma). In Zecca et al. [36] this split coincided in timing with our Node A. Using sequence data of a single chloroplast gene (rbcL) and no fossil calibrations, Xiang et al. [55] estimated the divergence time of 11 East Asia-North American disjunct species, representing 11 diverse genera of flowering plants at an average of 4.98 Ma and ranging from 5.98 Ma to <0.28 Ma. Finally, we estimated 9.78 Ma to 5.28 Ma for higher level divergences among the extant taxa in Asia, and 6.36 Ma (9.8 Ma –3.36 Ma) for the separation of Europe and the Near East from Asia (Node C in Figure 5), a separation that was estimated by Zecca et al. [36] at 4.31 Ma or 4.47 Ma (6.60 Ma -2.31 Ma, 6.71 Ma – 2.61 Ma). We estimated species-level divergences between 11.68 Ma (*V. nesbittiana*) and 1.3 Ma (*V. piasezkii, V. betulifolia;* and intraspecific divergences shown more fully in Figures 4 and 5 extended into more recent times. Overall, the timeframe estimated by the present study was more ancient than the estimates by other*Vitis*-specific molecular studies (Table1).

Table 1: Comparison of divergence estimates in *Vitis* among five studies that analyzed SNP data and used zero, two or three fossil calibration point's Continental origin, dispersal and diversification of *Vitis*

Node1	Xiang et al. 2000[2] [55]	Nie et al. 2010[3] [34]	Zecca et al. 2012[4,5] [36]	Liu et al. 2013[6][35]	Present study[7]
Vitis crown					
Estimate (Ma)	_[8]	~8	18.60, 19.05[5]	~12	28.32
95% HPD (Ma)	_	~16-4	28.79-9.50, 29.07-10.2	~22-6	41.25-16.23
subg. *Vitis* crown					
Estimate (Ma)	_	~5.5	6.31, 6.55	~7	17.82
95% HPD (Ma)	_	~10-2.5	9.50-3.50, 9.34-4.50	~11-5	26.71-10.14
Europe/Near East - Asia					
Estimate (Ma)	_	_	4.31, 4.47	_	6.36
95% HPD (Ma)	_	_	6.60-2.31, 6.71-2.61	_	9.80-3.36
Eurasia-NC America					
Estimate (Ma)	4.98 (5.98-<0.28)	_	6.31, 6.55	_	11.12

95% HPD (Ma)	–	–	9.50-3.50, 9.34-4.50	–	16.58-6.59
Higher level intra North America					
Estimate (Ma)	–	–	–	–	15.75-8.96
95% HPD (Ma)	–	–	–	–	multiple
Higher level intra Asia					
Estimate (Ma)	–	–	–	–	9.78-5.28
95% HPD (Ma)	–	–	–	–	multiple
Species stem					
Estimate (Ma)	–	–	–	–	11.68-1.3
95% HPD (Ma)	–	–	–	–	multiple
Fossil calibrations (Ma), Priors used					
Vitaceae - *Leea*	–	85±4.0[9]	85±4.0	90.7±1.0[10]	90.7±1.0
Vitis -Parthenocissus	–	58.5±5.0[11]	–	58.5±5.0	58.5±5.0
Old World - New world*Parthenocissus*	–	–	21.6±6.4[12]	–	–

Within *Vitis*	–	–	4.5 [13]	5.75±0.5[14]	5.75±0.5

[1]For nodes "Higher level intra North America", "Higher level intra Asia, "Species stem" the range contains several clades with different estimates, see Figure 3 and for details, and the 95% HPD included multiple HDP associated with the several clades, see Figure 3. [2]one chloroplast gene, 11 diverse Angiosperms; [3]4226 nt (2794 cpDNA+1432 nDNA); [4]3583 nt (2701 cpDNA+882 nDNA); [5]estimates from analyses that either constrained the stem (first number) or the crown of subgenus *Vitis*; [6]5471 nt cpDNA;[7]11,440 nt, 27 nuclear gene fragments; [8]dash: information is not applicable; [9]Wikström et al. [56] used a 3-gene data set, rate smoothing, and a single calibration point to estimate the stem age of Vitaceae at 78 - 92 Ma from which the prior distribution was derived; [10]estimate from Magallón and Castillo [57] because Liu et al. [35] agree with arguments made by Nie et al. [58] that the previous Vitaceae stem age estimate [56] was inaccurate; [11]stem age of the *Vitis-Ampelocissus* clade, located in the Paleocene to which the oldest reliable Vitaceae fossils date, including *Ampelopsis, Ampelocissus and Vitis*[35]; [12]prior based on estimates by Nie et al. [35]; [13]minimum age constraint based on fossil seeds described by Gong et al. [59]; [14]based on fossil relatives of *V. labrusca* described by Gong et al. [59].

Wan *et al.*

Wan *et al. BMC Evolutionary Biology* 2013 **13**:141, doi: 10.1186/1471-2148-13-141

Continental Origin, Dispersal and Diversification of *Vitis*

The phylogenetic relationships and network of grapevines reflect the Northern hemisphere Cenozoic history. The extensive ancestral reticulations revealed by the network and analysis of individual genes suggested well connected ancestral populations and species throughout the distribution followed by increasing range-wide fragmentation, isolation, and differentiation. The ancestral area analysis and the recurring distributional trend of American paraphyly with Eurasia in this study suggested a progression from North America to Asia to Europe consistent with previous studies [17,29,33,36]. However, Péros et al. [23] concluded that their analysis may support an Asian origin of *Vitis*. Fossils of Vitaceae have been found frequently in Western North American Eocene deposits (55.8 to 33.9 Ma) and have not yet been found in southeastern localities [60]. Fossils of *Vitis* seed were found in deposits of the Rocky Mountains and Great Plains of North America [34,53] and in central Europe [34,54]. These findings assigned the oldest age of *Vitis* to the Paleocene (65.5-58.8 Ma). At that time the supercontinent Laurasia had only begun dividing into North America and Eurasia [61] and the climate was considerably warmer in the northern latitudes [62]. These factors facilitated dispersal of warm-temperate terrestrial organisms in the northern hemisphere. Most East Asia–North American disjuncts from diverse families have had longer histories in North America than in Asia: of nine woody East Asian–East North American disjunct genera [60] all appeared earlier in the fossil record of North America than in that of Asia [63]. Wen et al. [64] found many more lineages with North American origins and migration to Asia than *vice versa*[58]. Nie et al. [58] argue for a North American origin of*Ampelopsis* (Vitaceae). Molecular phylogenetic analyses of several disjunct genera suggested a progression from East Asia to Eastern and Western North America [65,66]. Thus, the balance of grape-specific information tends to support our findings of a North American origin for the most recent common ancestor of *Vitis*.

After fragmentation of a Paleo/Neogene range, our phylogenetic trees suggested isolation of some North American and Asian species during the Plio- and Pleistocene cooling cycles, post glacial range expansions, and ecological adaptation. Much of the current Eastern North American range of *V. riparia*,*V. labrusca*, *V. aestivalis,* and *V. cinerea* was unsuitable for *Vitis* during the Wisconsinan glaciations due to coverage by the polar ice sheet and harsh conditions along its southern edge (Figure 1). These species must have expanded to their large present ranges after the glacial period. Large range expansions with post glacial warming were also promoted by the physiographic homogeneity of Eastern North America [63]. Fragmentation and local adaptations were evident in physiographically heterogeneous western North America and temperate eastern Asia. The North American species *V. shuttleworthii, V. nesbittiana, V. girdiana, V. palmata, V. bloodworthiana* and *V. blancoii* have smaller ranges and multiple species-specific SNP character changes. Similarly, physiographically diverse eastern Asia [63] had three species with multiple species-specific SNPs: *V. bashanica, V. hancockii, V. quinquangularis*. Local adaptations in heterogeneous environments likely lead to smaller population sizes and thus more rapid loss or fixation of novel characters [67].

The underlying evolutionary scenario for *Vitis* is consistent with origin in the Eocene, a time of maximum development of temperate

Paleo/Neogene forests. This was followed by diversification in the mid-Oligocene, the rise of subg. *Vitis* in the early Miocene, the North American and Asian disjunction in the late Miocene, range restriction and fragmentation and speciation during the Pliocene and Pleistocene cooling cycles. These caused the primary divisions within *Vitis* as well as species-level and some intra-specific divisions [68]. The North Atlantic land bridge was present in the early Paleogene[69,70] and may have no longer existed when *Vitis* arose, leaving Beringia as the major route for potential gene flow. The area of the Bering and Chukchi seas lay above sea level for most of the last 50 to 60 M years [71] and was suitable for exchanges of temperate plants [69] until the establishment of the Bering Seaway 3.5-5 Ma [72], permitting genetic exchange at least until late Miocene to which the disjunction was timed. The Pleiocene/Pleistocene cooling cycles are well known to have caused range restrictions, survival in refugia, and diversifications in many groups of organisms [73], both on land and in the sea. This study shows clearly that *Vitis* was also a part of this great biogeographic phenomenon.

Phylogenetically-Based *Vitis* Systematics

The systematics of *Vitis* is a challenging area of taxonomy. Our findings confirmed the tenuous nature of many grapevine species and especially higher groupings such as series. The apparent species-specific SNPs are good candidates to apply in species delineation investigations of grapes.

The present study found very low support for all series that included more than one species except for the Munson/Moore series Precoces/Ripariae (Figure 4, node W). Other well supported higher-level groupings were subg. *Vitis* (Figure 4, node A) and genus *Vitis*, supporting the division of the genus*Vitis* into two sections [4,74,75] or subgenera [5,76]. lists a synopsis of the major*Vitis* classifications. Only Galet [4] assigned Asian species to series. The most comprehensive treatment of Chinese *Vitis*[6] did not apply a series-level classification. Most Chinese species could be assigned to one series if series were to be used (Figure 5). This may not include *V. amurensis*, *V. coignetiae* and *V. yenshanensis* as these species in some analyses grouped firmly within Asia (as opposed to Figure 5where they are basal to Eurasia). It appears as if *V. jacquemontii* should be assigned to the series*Viniferae*. However, our accessions had perfect flowers, suggesting past hybridization with

V. vinifera. The phylogenetic position intermediate between the Asian and Eurasian species and the well-defined split revealed in the network (Figure 2) supported this conclusion.

The derived position of *V. sylvestris* was unexpected. *V. sylvestris* is the suggested progenitor of *V. vinifera*[77] while the phylogenetic position suggests that *V. sylvestris* was derived from *V. vinifera*(Figure 5). This may be an artifact of the tenuous nature of most higher-level relationships revealed in this study. It could also be a result of the nature of selection and clonal propagation that all *V. vinifera*cultivars included in the present study have been subjected to, some of them potentially for thousands of years [77,78]. Evolution is arrested by clonal propagation, leaving the naturally evolving wild species to appear more derived. Myles et al. [21] concluded that current commercial *V. vinifera* varieties are only one or two generations removed from the wild *V. sylvestris*.

Mullins et al. [10] hypothesized Asian/North American sister species pairs for *V. coignetiae/V. labrusca*and for *V. jaquemontii/V. tiliifolia* (*V. lanata* and *V. caribaea* in [10]). Our results did not support sister pair relationships for *V. coignetiae /V. labrusca* as these species placed solidly into well separated Asian and North American clades, respectively. Our results are inconclusive with respect to the *V. jacquemontii/V. tiliifolia* pair due to the possible hybrid nature of *V. tiliifolia* accessions in general and the dispersed positions of *V. tiliifolia* samples.

V. girdiana has been considered to be a variety of *V. arizonica*[75], a variety of *V. californica*[79] and its own separate species [75]. Our results preliminarily identified *V. girdiana* as a well supported independent species (using the general lineage concept [80] and diagnosability [81]) with five species-specific SNPs. More samples need to be investigated to assess the discriminatory power of these SNPs. Wada [31] also identified a monophyletic *V. girdiana* cluster, although it had poor bootstrap support.

Samples 080-084 came to us as *V. cinerea* (Engelm.) Engelm. ex Millardet var. *floridana* (Munson) but placed solidly into the *V. aestivalis* clade. This highlighted confusion in the past related to the synonym*Vitis simpsonii* that has been claimed for two different species as described in Comeaux [82], one belonging to Aestivales and the other to Cinerescentes. The synonym *V. rufotomentosa* has the same problem.

Our study showed conclusively that these accessions belong to *V. aestivalis*. Several additional accessions were identified as misnamed and others were recognized as hybrids. Finally, two accessions, 111*V. flexuosa* DVIT1385 and 304*V. wilsoniae* Wangmaiputao, were of Asian origin yet grouped with North American accessions and remain anomalies that could not be resolved.

CONCLUSIONS

This is the first study to apply sequences of a large number of nuclear loci combined with extensive species and intraspecific sampling to the phylogeny and biogeographic history of *Vitis* and the problem of *Vitis* systematics. The genome-wide sampling of SNPs provided insight into the evolutionary history of the grape genus and supported previous notions of Paleogene origins, range fragmentation, and recent nature of the species, joining *Vitis* with the large group of organisms whose extant species differentiated in response to Pliocene and Quarternary climate change [73]. We found that the most recent common ancestor of *Vitis* was North American. The major clades formed throughout the native distribution at 23-8 Ma (broad range due to large HPDs), suggesting that vicariance (the fragmentation of a large Paleo/Neogene Northern hemisphere distribution) in conjunction with local adaptation, was a dominant force in structuring genetic diversity of extant *Vitis spp.* We demonstrated that genome-wide nuclear SNPs were a productive approach to address questions at and below the species level in grapes. Many species were well supported, and the markers with low homoplasy defining those lineages will likely be useful in species delineation and assessing the reliability of different morphological taxonomic characters. Most higher-level relationships within the genus suffered from weak support. The genus itself was extremely well supported. This suggested that the phylogenetic signal was too weak to overcome the level of noise created by evolutionary forces acting within the *Vitis* gene pool. Two of the most important forces, probably acting concurrently or alternating, are incomplete lineage sorting of ancestral polymorphism and reticulation. Broad reticulation across many species probably prevented the ancestral gene pool from diverging during the Neogene forest stage, maintained reproductive compatibility, and is still acting today as evidenced by the prevalence of hybrids found in the wild and in repository collections. However, climatic oscillations during the Pliocene and Quaternary, coupled with physiographic heterogeneity, provided enough recent barriers to gene flow to facilitate evolutionary divergence. In light of the recency of divergence and diffuse genetic boundaries, higher-level taxonomic groupings, such as series, may be misleading.

METHODS

Plant Materials

A total of 309 accessions of 48 species or varieties (~80% of the approximately 60 known species of the genus) and outgroups were sampled in this study: 21 species from Asia, both European species, and 25 species and varieties from North America (Figure 1; These samples were obtained from: 1) the Grape Germplasm Collection at the Northwest A&F University (NAFU), Yangling, Shaanxi Province, China, (DNA), 2) USDA-ARS, Plant Genetic Resources Unit (PGRU), Geneva, NY, USA, and 3) USDA-ARS, National Clonal Germplasm Repository (NCGR), Davis, CA, USA. Four closely related genera based on chloroplast and nuclear markers [19,28,32,34] were chosen as outgroups in the dating of divergences using BEAST: *Parthenocissus* spp. Planch, *Ampelopsis glandulosa* (Wall.) Momiy. var. *brevipedunculata* (Maxim.) Momiy, *Leea coccinea* Planch. 'Rubra', *Cayratia japonica* Thunb. Two of the outgroup genera (five species of *Cissus* and *Cayratia japonica*) were obtained from a research collection (Dr. P. Cousins, USDA/ARS, presently E & J Gallo Winery). The outgroup *Leea coccinea*"Rubra" was grown from seeds obtained from Carter Seeds (Vista, California).

No cultivars of *Vitis spp.* were included except for *V. vinifera* ssp. *vinifera* for which no wild accessions are known [77]. To mitigate long-branch attraction, the 40 chromosome *Vitis rotundifolia* Michx. subg.*Muscadinia* was used as the outgroup in analyzing subg. *Vitis*. This was justified by all preliminary analyses on the complete data set that identified *V. rotundifolia* as the sister species to subgenus *Vitis*and it is consistent with other studies e.g. [23,29,33-36]. One to 27 accessions or genotypes were sampled per species. All available varieties (not cultivars) of a species were sampled. Similarly, widely distributed species were more extensively sampled to include potential geographic differentiation. A few sibling groups were also included to test the ability of the markers to place or distinguish those accessions.

Based on preliminary analyses, accessions that placed in unexpected positions or had very weak support on preliminary phylogenetic trees were submitted to a taxonomic expert (Dr. P. Cousins) for an independent assessment of species identity, but without indicating the nature of the conflict. lists the results of all assessments. The labels in the figures and tables indicate the corrected names unless otherwise indicated. Exact geographic coordinates of origin were not available for most accessions. Accessions and pertinent details are listed in. The accessions located in the US repositories can be requested through the Genetic Resources Information Network (GRIN) [83] and plant materials (leaves, cuttings) can be requested from the clonally maintained vines at these sites.

DNA Isolation and Re-Sequencing

DNA was isolated from fresh or frozen young leaves and apical meristems using a modified CTAB (cetyltrimethylammonium bromide; Sigma H6292) protocol [84,85] with 2-5% PVP (Polyvinylpyrrolidone, mol. wt. 40,000; Sigma PVP40) in the extraction buffer to remove secondary compounds, two chloroform purifications to remove proteins and a NaCl and ethanol precipitation to remove polysaccharides.

Primer screening was performed in 25 μL PCR volumes. 50 μL PCR volumes were cleaned for sequencing, concentrated and used in 12 μL cycle sequencing reactions. Provides the detailed conditions.

The exploratory sequencing was performed in-house at PGRU on ABI-3100xl Genetic Analyzer. The high-throughput sequencing (30 gene fragments for 309 accessions) was performed by Genaissance Pharmaceuticals, Inc (New Haven CT, USA). Both strands were separately sequenced using the PCR forward or reverse primer.

SNP Discovery and Selection

Expressed Sequence Tags (ESTs) [86] of *Vitis vinifera* and grape mRNAs in NCBI in 2004 were sub-clustered and surveyed to predict SNPs using an in-house pipeline as described by Labate and Baldo[87]. The 62 variably-sized EST libraries and additional grape mRNAs included 108,429 *V. vinifera*sequences which formed 3,792 clusters. Because EST data are often based on one sequencing pass and are not filtered for error, a predicted SNP may not be verifiable. Because this study was intended to survey broadly across the entire genus, gene markers that were predicted to be monomorphic among *V. vinifera* were discarded. Markers with extreme levels of polymorphism were also excluded to minimize possible selection of duplicated loci.

Pairs of PCR primers were designed using the program 'Primer 3' [88] for 281 gene fragments of 400-600 base pairs (bp) containing moderate polymorphism. The amplifications were tested using three DNA samples, one each from Asia (*V. romanetii* Rom. Caill. 'Jiangxi2'), Europe (*V. vinifera* 'Rotberger', DVIT2339) and North America (*V. rotundifolia*, DVIT1689). Robust, single bands were obtained for 201 of 281 primer pairs (71.7%). Then 96 primer pairs with robust single bands were chosen for re-sequencing to test sequence quality using eight species (*V. cinerea* (PI588575), *V. labrusca* L. (PI588194), *V. amurensis* (Zuoshan1), *V. quinquangularis* (Weinan3), *V. romanetii* (Pingli7), *V. davidii*(Xuefeng), *V. hancockii* (Lingye_F) and *V. yenshanensis* (Yanshan_F). Thirty of the most consistently amplifiable gene fragments, both within *Vitis* and outgroups, with suitable polymorphisms and only minor sequence length variation in the eight tested, were re-sequenced in a total of 309 accessions. Predicted genes were identified in comparison with the NCBI non-redundant protein sequence database. When the *V. vinifera* genome sequence became available [9], the primer sequences and gene fragments were BLASTed [89] against this genome to determine their chromosome locations and confirm their homology and identity. When the final dataset was assembled, three gene markers were excluded because of unalignable indels (one marker), and suspected duplicate loci (two markers). The sequences of the 27 final primer pairs and supporting information are listed in.

Sequence Alignment, Data Sets, Coding of Gaps

ProSeq [90] was used for editing sequence based on trace files, and Mutation Surveyor (Soft Genetics) was used for base calling. Heterozygotes were manually edited to use the IUPAC-IUB symbols for nucleotide nomenclature [91]. The results from ProSeq and Mutation Surveyor were compared for accuracy, and nearly 100% agreement was found. Discrepancies were resolved by examining trace files manually. Sequences were aligned manually and also aligned using Clustal W [92] with default parameters.

Extensive preliminary phylogenetic and PCA [93] and subsets of OTUs revealed known and new hybrids which were excluded in the final analysis because phylogenetic trees can be strongly influenced by hybrid taxa [94]. This does not guarantee that the final analyses were devoid of hybrids as they are not always identifiable based on morphology. The final phylogenetic data set contained 273 OTU composed of subgenera *Vitis* and *Muscadinia*. This dataset was modified for dating divergences using BEAST: 1) Four outgroup taxa were added to match calibration points in Nei et al. [34]: 060_*Ampelopsis brevipedunculata*, 096_*Leea coccinea* 'Rubra', 129_*Cayratia japonica*, 247_*Parthenocissus* spp.. *Leea* and *Cayratia* had substantial amounts of missing data With the presence of multiple individuals per species dating is a more complex issue and would preferably apply coalescence methods. Preliminary analyses indicated that the present data set was not sufficiently informative to allow a well-supported coalescent analysis. Thus, the number of ingroup OTUs was reduced to one accession per species and variety for efficient calculations[95] and to satisfy the Yule model of speciation. These modifications resulted in the 52-OTU dataset summarizes the members of each analysis. Preliminary ML analyses were conducted on all single gene fragments, and partitioned and unpartitioned concatenated sets for a total evidence dataset [96].

Gaps were treated as characters using 'simple indel coding' (SIC) [97] and implemented in SeqState 1.4.1 [98]. Simple gap coding was chosen because it is a preferred coding method for empirical studies[99,100]. Inclusion of gaps in phylogenetic analyses is limited by the optimality criterion used for phylogenetic inference. Gap information was used for parsimony analysis only. ML and BA treated gaps as missing data. Combinability of DNA partitions was ascertained using Wien's [101] method: existence of corresponding but incongruent clades with bootstrap support greater than 70% are seen as support for not concatenating data sets.

Test for Recombination, Network Analyses

The Phi Test [43] implemented in Splitstree4 [102] was used to test for recombination in each gene fragment.

The best single gene tree from each locus was combined into a file from which a consensus network was constructed in Splitstree4 [102]. Thresholds used were 0.04 (all splits present in at least one tree, 1/26), 0.08 (splits present in at least two of the trees), 0.5 (splits present in half the trees), 0.9 (splits present in 90% of the trees).

Neighbor Net with uncorrected P distance in SplitsTree4 [102,103] was used to visualize conflict in the 273-OTU matrix with 27 concatenated gene fragments.

Calculation of Divergence Time

The geologic time scale of Gradstein et al. [104] was used in this study. The term 'Tertiary' was replaced by Paleogene and Neogene [105].

Bayesian (BA) estimates in the BEAST V1.7.4 [95,106] software were used to estimate divergence dates using Markov Chain Monte Carlo (MCMC) sampling. Trees were visualized in Fig tree V1.3.1 [107]. Many preliminary runs were conducted on the partitioned and unpartitioned data file to explore parameters. Operators were optimized automatically. The final .xml files were run ten times using the maximum time available at the Computational Biology Service Unit (CBSU) BioHPC computer cluster at Cornell University. The conditions for the partitioned runs were: 27 unlinked partitions with individual substitution models, estimated frequencies of nucleotides, a random starting tree, an uncorrelated relaxed clock with log normally distributed uncorrelated rates between branches [108], Yule model of speciation (a pure birth process), default operators modified based on preliminary runs, auto-optimization turned on, and parameters were sampled every 10,000 steps. The conditions for the unpartitioned runs were identical except that there was only one partition and one substitution model. Marker-specific and whole-dataset-specific substitution models were determined in Findmodel [109] and are listed in. Findmodel uses Weighbor [110], PAML [111] and methods in Modeltest [112] to determine substitution models. Following the reasoning of Nie et al.[34] a normal prior was used with a mean of 58.5 Ma (st. dev. = 5 Ma) for the stem age of *Vitis*. Liu et al.'s [35] reasoning was adopted to 1) place the second calibration point of *V. labrusca* and closely related North American relatives in the subg. *Vitis* at 5.75 Ma (st. dev.=0.5 Ma), and 2) fix the stem age of *Vitaceae* with a normal prior distribution of 90.7 Ma (st. dev. = 1 Ma). Runs from the same .xml file were combined if they shared the same trace, met general quality requirements outlined in, and if the addition increased the ESS of key parameters. The three unpartitioned runs with the highest (almost identical) likelihoods were combined after 10% burnin was removed to produce the chronogram in Figure 3. Unpartitioned runs were used because the combined partitioned runs did not have sufficient support for important parameters. Five runs of the partitioned dataset were combined after removing a burnin of 10-75% to compute the evolutionary rates (gene.meanRate) of the individual genes. These gene.meanRates had large ESS. Means and 95% Highest Posterior Density (HPD) from these combined runs were computed using TreeAnnotator.

Phylogenetic and Ancestral Area Analyses

To evaluate species we used a phylogenetically based general lineage concept, where species are defined as separately evolving segments of metapopulation lineages [80]. Additional criteria of more stringent species definitions were considered, such as monophyly [113] and diagnosability [81,114].

Maximum Likelihood

The 273-OTU matrix was analyzed under the ML criterion using RAxML [115] versions 7.0.4, 7.2.6 and HPC2 at the web portal of the Cyber Infrastructure for Phylogenetic Research (CIPRES) cluster. The data were partitioned by gene fragment. All characters were included. Gap coding was removed. Indels were treated as missing data. Twenty replicate searches were run on this final data set using a rapid hill climbing algorithm and the GTRGAMMA (= GTR + Optimization of substitution rates + GAMMA model of rate heterogeneity, the alpha parameter was estimated) model of substitution as recommended by the program's author, and the default of 25 rate categories. The rapid bootstrapping option [116] was chosen to generate 1,000 bootstrap replicates. The best-scoring ML tree was obtained in the same search and bootstrap values were annotated. Output was visualized in Dendroscope V2.2.2 [117].

To test for conflict among genes, a preliminary analysis was performed for each gene fragment using the same parameters as above but without a partitioned model. Each gene was run in 10 replicates. Five more replicates were added if the final maximum likelihood values varied extensively among replicates. To keep trees comparable no OTUs were excluded. One-thousand bootstrap replicates were collected for each gene fragment. Incongruent clades with bootstrap support of 70% or greater were considered as support for not combining data sets [42,100]. Due to computational limitations, the single gene analyses were performed only in RAxML at CIPRES. Because of the sparse yield of information and low information content of most markers, this computational expense was not repeated with the 273-OTU matrix.

Bayesian (BA)

Bayesian analyses were performed on the 273-OTU matrix using Mr. Bayes [118] on the concatenated, non-partitioned data set using the K80 substitution model (Nst=2, 4 by 4) plus Gamma, as determined in Findmodel. Multiple short runs were performed to determine the temperature and number of chains that would support chain swap. The two final and longest runs of 48 and 50 million generations (fitting just within the 168 hr time limit) were run with 8 chains, temperature of 0.10, and sample frequency every 5,000 generations. Tracer V1.5 [119] was used to evaluate the MCMC runs, TreeAnnotator v1.6.1 [95] and Figtree v1.3.1 were used to annotate and visualize maximum credibility trees listing all posterior probabilities. Burn in was 2.5 million (50 mil run) and 10 million (48 mil run).

Maximum Parsimony (MP)

The software package TNT, Hennig Society version [120] was used to analyze the unpartitioned 273-OTU matrix under parsimony, assuming unordered character state transformation and equal weights (Fitch parsimony) [121]. Uninformative characters were excluded and gaps were coded. The efficient option of "driven search" in TNT was used for the search. This option searched until a minimum tree length was found a certain number of times and then a consensus was estimated. After a second round of searching the new consensus was compared to the previous one, and so on until the consensus stabilized [122]. The driven searches included the ratchet [123], tree drifting, tree fusion and sectorial search [124]. The default settings were used except that the consensus was stabilized four times instead of twice. The search was repeated four times using a different random starting seed and without specifying a target score. The strict consensus tree was constructed using all most parsimonious trees from all four searches. Bootstrap support was based on 1,000 replicates of driven searches using the same search components and default parameters. Synapomorphies were optimized and listed, and those of interest were reconstructed. The character values, indicating the level of homoplasy, were obtained in TNT to study the support at nodes of interest. Trees for illustrations were exported in a NEXUS format, manually converted to Newick trees, visualized and annotated using Dendroscope V 2.2.2.

Ancestral Area Analysis

The geographic distribution was partitioned into four continental area units that correspond to broad distributional trends in *Vitis*: Asia, Europe/near East, E- and SE-North America (including Mexico and the Caribbean) and Western North America (Figure 1). An area code was added to each accession in the 273-OTU data matrix. Fitch optimization (reversible parsimony) [121] was performed in TNT to optimize the area on the strict consensus tree [125].

AUTHORS' CONTRIBUTIONS

YW, CJS, AMB, HRS conceived the ideas; YW, CJS collected or arranged for the materials; AMB, YW, JAL developed the markers; YW, AMB developed the sequence matrix; HRS analyzed the data except for PCA; AMB developed and performed PCA; HRS, JAL, YW wrote the paper; CJS and GYZ managed the project. All authors read and approved the final submission.

ACKNOWLEDGEMENTS

We would like to thank Ashley Egan for supplying crucial sparks and expertise at several analytical key points of this project. Warren Lamboy worked extensively on data accuracy, management, and early analysis of datasets. Pablo Goloboff answered many basic and operational questions about TNT. Mark Miller and Wayne Pfeiffer among others at the San Diego Supercomputer Center gave very effective support to RAxML and endeavored to bring BEAST online at the CIPRES portal. Lauren Chan provided answers to specific questions in BEAST and Mr. Bayes. Peter Cousins supplied outgroup material and offered excellent advice on grape species identification and anything else related to grapes. Bob Nearpass provided local IT support, NCGR in Davis, CA supplied tissue samples. Anonymous reviewers provided suggestions that improved this work. Part of this work was carried out using the resources of the Computational Biology Service Unit from Cornell University which is partially funded by Microsoft Corporation. USDA is an equal opportunity provider and employer.

REFERENCES

1. Food and Agriculture Organization of the United Nations: *FAO*. [http://faostat.fao.org
2. Reisch BI, Pratt C: Grapes. In *Fruit breeding*. 2nd edition. Edited by Janick J, Moore JN. New York: Wiley; 1996:297-369.
3. Comeaux BL, Nesbitt WB, Fantz PR: Taxonomy of the native grapes of North Carolina. *Castanea* 1987, 52:197-215.

4. Galet P: *Cepages et vignobles de France. Tome 1 Les vignes Americaines*. Dehan: Montpellier; 1988.
5. Moore MO: Classification and systematics of Eastern North American *Vitis* L. (Vitaceae) north of Mexico. *Sida* 1991, 14:339-367.
6. Chen Z, Hui R, Wen J: Vitaceae. *Flora of China* 2007, 12:173-222] PDF at [http://flora.huh.harvard.edu/china/mss/volume12/Vitaceae.pdf
7. Alleweldt G, Spiegel-Roy P, Reisch B: Grapes (*Vitis*). In *Genetic resources of temperate fruit and nut crops*. Edited by Moore JN, Ballington JR. Wageningen: Acta Hort 290; 1990:291-337.
8. Patel GI, Olmo HP: Cytogenetics of Vitis: I. The hybrid *V. vinifera* x *V. rotundifolia*. *Am J Bot* 1955, 42:141-159.
9. Jaillon O, Aury J-M, Noel B, Policriti A, Clepet C, Casagrande A, Choisne N, Aubourg S, Vitulo N, Jubin C, Vezzi A, Legeai F, Hugueney P, Dasilva C, Horner D, Mica E, Jublot D, Poulain J, Bruyère C, Billault A, Segurens B, Gouyvenoux M, Ugarte E, Cattonaro F, Anthouard V, Vico V, Del Fabbro C, Alaux M, Di Gaspero G, Dumas V, *et al.*: The grapevine genome sequence suggests ancestral hexaploidization in major angiosperm phyla. *Nature* 2007, 449:463-467.
10. Mullins MG, Bouquet A, Williams LE: The grapevine and its relatives. In *Biology of the grapevine*. Edited by Mullins MG, Bouquet A, Williams LE. Cambridge: Cambridge University Press; 1992:17-36.
11. Comeaux BL: *Taxonomic studies on certain native grapes of the Eastern United States. PhD thesis*. North Carolina State University: Department of Horticultural Science; 1984.
12. Rogers DJ, Rogers CF: Systematics of North American grape species. *Am J Enol Vitic* 1978, 29:73-78.
13. Barrett HC, Cramer SG, Rhodes AM: A taximetric study of interspecific variation in *Vitis*. *Vitis* 1969, 8:177-187.
14. Reinig WF: *Die Holarktis*. Jena: Gustav Fischer Verlag; 1937.
15. Arnold C, Rosetto M, McNally J, Henry RJ: The application of SSRs characterized for grape (*Vitis vinifera*) to conservation studies in Vitaceae. *Am J Bot* 2002, 89:22-28.

16. Aradhya MK, Dangle G, Prins BH, Boursiquot J- M, Walker MA, Meredith CP, Simon CJ: Genetic structure and differentiation in cultivated grape, *Vitis vinifera* L. *Genet Res, Camb* 2003, 81:179-192.
17. Aradhya MK, Koehmstedt A, Prins BH, Dangle GS, Stover E: Genetic structure, differentiation, and phylogeny of the genus *Vitis*: implications for genetic conservation. *Acta Hort* 2008, 799:43-49.
18. Grassi F, Labra M, Imazio S, Ocete-Rubio R, Failla O, Scienza A, Sala F: Phylogeographical structure and conservation genetics of wild grapevine. *Conserv Genet* 2006, 7:837-845.
19. Ingrouille MJ, Chase MW, Fay MF, Bowman D, Van der Bank M, Bruijn ADE: Systematics of Vitaceae from the viewpoint of plastid *rbcl* DNA sequence data. *Biol J Linn Soc* 2002, 138:421-432.
20. Lijavetzky D, Cabezas JA, Ibanez A, Rodriguez V, Martinez-Zapater JM: High throughput SNPdiscovery and genotyping in grapevine (*Vitis vinifera* L.) by combining a re-sequencing approach and SNPlex technology. *BMC Genomics* 2007, 8:424.
21. Myles S, Boyko AR, Owens C, Brown PJ, Grassi F, Aradhya MK, Prins B, Reynolds A, Chia J-M, Ware D, Bustamante CD, Buckler ES: Genetic structure and domestication history of the grape. *Proc Natl Acad Sci* 2011, 108:3530-3535.
22. Myles S, Chia J-M, Hurwitz B, Simon C, Zhong GY, Buckler E, Ware D: Rapid genomic characterization of the genus *Vitis*. *PLoS One* 2010, 5(e8219):1-9.
23. Péros J-P, Berger G, Portemont A, Boursiquot J-M, Lacomb T: Genetic variation and biogeography of the disjunct *Vitis* subg. *Vitis* (Vitaceae). *J Biogeogr* 2011, 38:471-486.
24. Arroyo-Garcia R, Ruiz-Garcia L, Bolling L, Ocete R, Lopez MA, Arnold C, Ergul A, Soylemezoglu G, Uzun HI, Cabello F, Ibanez J, Aradhya MK, Atanassov A, Atanassov I, Balint S, Cenis JL, Costantini L, Goris-Lavets S, Grando MS, Klein B, McGovern PE, Merdinoglu D, Pejic I, Pelsi F, Primikirios N, Risovannaya V, Roubelakis-Angelakis KA, Snoussi H, Sotiri P, Tamhankar S, *et al.*:Multiple origins of cultivated grapevine (*Vitis vinifera* L. ssp. *sativa*) based on chloroplast DNA polymorphisms. *Mol Ecol* 2006, 15:3707-3714.

25. Rossetto M, Crayn DM, Jackes BR, Porter C: An updated estimate of intergeneric phylogenetic relationships in the Australian Vitaceae. *Can J Bot* 2007, 85:722-730.

26. Rossetto M, McNally J, Henry RJ: Evaluating the potential of SSR flanking regions for examining taxonomic relationships in the Vitaceae. *Theor Appl Genet* 2002, 104:61-66.

27. Salmaso M, Faes G, Segala C, Stefanini M, Salakhutdinov I, Zyprian E, Toepfer R, Grando MS, Velasco R: Genome diversity and gene haplotypes in the grapevine (*Vitis vinifera* L.), as revealed by single nucleotide polymorphisms. *Mol Breed* 2004, 14:385-395.

28. Soejima A, Wen J: Phylogenetic analysis of the grape family (Vitaceae) based on three chloroplast markers. *Am J Bot* 2006, 93:278-287.

29. Tröndle D, Schröder S, Kassemeyer H-H, Kiefer C, Koch MA, Nick P: Molecular phylogeny of the genus *Vitis* (Vitaceae) based on plastid markers. *Am J Bot* 2010, 97:1168-1178.

30. Vezzulli S, Micheletti D, Riaz S, Pindo M, Viola R, This P, Walker MA, Troggio M, Velasco R: A SNP transferability survey within the genus *Vitis*. *BMC Plant Biol* 2008, 8:128.

31. Wada EB: *Systematics and evolution of Vitis. PhD thesis.* Ann Arbor: ProQuest LLC; 2008. [*UC Davis, Plant Biology*] Available through Proquest UMI Microform 3336358

32. Wen J, Nie Z-L, Soejima A, Meng Y: Phylogeny of Vitaceae based on the nuclear *GAI1*gene sequences. *Can J Bot* 2007, 85:731-745.

33. Aradhya M, Wang Y, Walker MA, Prins BH, Koehmstedt AM, Velasco D, Gerrath JM, Dangle GS, Preece JE: Genetic diversity, structure, and patterns of differentiation in the genus*Vitis*. *Plant Syst Evol* 2013, 299:317-330.

34. Nie Z-L, Sun H, Chen Z-D, Meng Y, Manchester SR, Wen J: Molecular phylogeny and biogeographic diversification of *Parthenocissus* (Vitaceae) disjunct between Asia and North America. *Am J Bot* 2010, 97:1342-1353.

35. Liu X-Q, Ickert-Bond SM, Chen L-Q, Wen J: Molecular phylogeny of *Cissus* L. of Vitaceae (the grape family) and evolution of its

pantropical intercontinental disjunctions. *Mol Phylogenet Evol* 2013, 66:43-53.

36. Zecca G, Abbott R, Sun W-B, Spada A, Sala F, Grassi F: The timing and mode of evolution of wild grapes (*Vitis*). *Mol Phylogenet Evol* 2012, 62:736-747.
37. Garcia-Pereira MJ, Caballero A, Quesada H: Evaluating the relationship between evolutionary divergence and phylogenetic accuracy in AFLP data sets. *Mol Biol Evol* 2010, 27:988-1000.
38. Delsuc F, Brinkmann H, Philippe H: Phylogenomics and the reconstruction of the tree of life. *Nat Rev Genet* 2005, 6:361-375
39. Burleigh JG, Bansal MS, Eulenstein O, Hartmann S, Wehe A, Vision TJ: Genome-scale phylogenetics: inferring the plant tree of life from 18,896 gene trees. *Syst Biol* 2011, 60:117-125.
40. Maureira-Butler IJ, Pfeil BE, Muangprom A, Osborn TC, Doyle JJ: The reticulate history of*Medicago* (Fabaceae). *Syst Biol* 2008, 57:466-482.
41. Doyle JJ, Doyle JL, Brown AHD: Incongruence in the diploid B-genome species complex of*Glycine* (leguminosae) revisited: Histone H3-D alleles versus chloroplast haplotypes. *Mol Biol Evol* 1999, 16:354-362.
42. Sang T, Zhong Y: Testing hybridization hypotheses based on incongruent gene trees. *Syst Biol* 2000, 49:422-434.
43. Bruen TC, Philippe H, Bryant D: A simple and robust statistical test for detecting the presence of recombination. *Genetics* 2006, 172:2665-2681.
44. Lanier HC, Knowles LL: Is recombination a problem for species-tree analyses? *Syst Biol* 2012, 61:691-701.
45. Huson DH, Bryant D: Application of Phylogenetic Networks in Evolutionary Studies. *Mol Biol Evol* 2006, 23:254-267 Software at [http://www.splitstree.org
46. Rokas A, Carroll SB: Bushes in the tree of life. *PLoS Biol* 2006, 4:1899-1904.
47. Brandley MC, Warren DL, Leache AD, McGuire JA: Homoplasy and clade support.*Syst Biol* 2009, 58:184-198.
48. Wolfe KH, Li W-H, Sharp PM: Rates of nucleotide substitution vary greatly among plant mitochondrial, chloroplast, and nuclear DNA.*Proc Natl Acad Sci USA* 1987, 84:9054-9058.

49. Degnan JH, Rosenberg NA: Discordance of species trees with their most likely gene trees.*PLoS Genet* 2006, 2(e68):62-768.

50. Wiens JJ: Polymorphism in systematic and comparative biology. *Annu Rev Ecol Syst* 1999, 30:327-362.

51. This P, Lacomb T, Cadle-Davidson M, and Owens C: Wine grape (*Vitis vinifera* L.) color associates with allelic variation in the domestication gene *VvmybA1.Theor Appl Genet* 2007, 114:723-730.

52. Philippe H, Brinkmann H, Lavrov DV, Littlewood DTJ, Manuel M, Wörheide G, Baurein D:Resolving Difficult Phylogenetic Questions: Why More Sequences Are Not Enough. *PLoS Biol* 2011, 9:e1000602.

53. Brown RW: Paleocene flora of the Rocky Mountains and Great Plains. *US Geol Surv Prof Pap* 1962, 375:1-120.

54. Mai DH: Neue früchte und samen aus Paläozänen ablagerungen Mitteleuropas. *Feddes Repertorium* 1987, 98:197-229.

55. Xiang Q-Y, Soltis DE, Soltis PS, Manchester SR, Crawford DJ: Timing of the Eastern Asian-Eastern North American floristic disjunction: molecular clock corroborates paleontological estimates. *Mol Phylogenet Evol* 2000, 15:462-472.

56. Wikström N, Savolainen V, Chase MW: Evolution of the Angiosperms: calibrating the family tree. *Proc R Soc Lond B* 2001, 268:2211-2220.

57. Magallón S, Castillo A: Angiosperm diversification through time. *Am J Bot* 2009, 96:349-365.

58. Nie Z-L, Sun H, Manchester SR, Meng Y, Luke Q, Wen J: Evolution of the intercontinental disjunctions in six continents in the Ampelopsis clade of the grape family (Vitaceae). *BMC Evol Biol* 2012, 12:17. http://www.biomedcentral.com/1471-2148/12/17.

59. Gong F, Karsai I, Liu Y-S: Vitis Seeds (Vitaceae) from the late Neogene Gray fossil site, Northeastern Tennessee, U.S.A. *Rev Palaeobot Palyno* 2010, 162:71-83.

60. Manchester SR: Biogeographical relationships of North American Tertiary floras. *Ann Missouri Bot Gard* 1999, 86:472-522

61. Press F, Siever R: *Earth*. 4th edition. New York: W.H. Freeman and Company; 1985.

62. Bowen GJ, Beerling DJ, Koch PL, Zachos JC, Quattlebaum T: A humid climate state during the Palaeocene/Eocene thermal maximum. *Nature* 2004, 432:495-499.
63. Qian H, Ricklefs RE: Large-scale processes and the Asian bias in species diversity of temperate plants. *Nature* 2000, 407:180-182.
64. Wen J, Ickert-Bond S, Nie Z-L, Li R: Timing and modes of evolution of eastern Asian- North American biogeographic disjunctions in seed plants. In *Darwin's Heritage Today: Proceedings of the Darwin 2010 Beijing International Conference*. Edited by Long M, Gu H, Zhou Z. Beijing: Higher Education Press; 2010:252-269.
65. Tiffney BH, Manchester SR: The use of geological and paleontological evidence in evaluating plant phylogeographic hypotheses in the northern hemisphere Tertiary.*Int J Plant Sci* 2001, 162(Suppl 6):S3-S17.
66. Xiang Q-Y, Soltis DE, Soltis PS: The Eastern Asian and Eastern and Western North American floristic disjunction: congruent phylogenetic patterns in seven diverse genera. *Mol Phylogenet Evol* 1998, 10:178-190.
67. Hartl DL, Clark AG: *Principles of population genetics*. Sunderland: Sinaur; 1989.
68. DeLattin G: **Über** den Ursprung und die Verbreitung der Reben. *Der Züchter* 1939, 11:217-225.
69. Wen J: Evolution of East Asian and Eastern North American disjunct distributions in flowering plants. *Annu Rev Ecol Syst* 1999, 30:421-455.
70. Tiffney BH: The Eocene North Atlantic land bridge: its importance in Tertiary and modern phytogeography of the northern hemisphere. *J Arnold Arbor* 1985, 66:243 73.
71. Hopkins DM: Cenozoic history of the Bering land bridge. *Science* 1959, 129:1519-1528.
72. Marincovich L, Gladenkov AY: Evidence for an early opening of the Bering Strait. *Nature* 1999, 397:149-151.
73. Hewitt GM: The genetic legacy of the Quaternary ice ages. *Nature* 2000, 405:907-913.
74. Planchon JE: Monographie des Ampelidae varies. *Monographia Phanerogamerum* 1887, 5:305-368.

75. Munson TV: *Foundations of American grape culture*. Denison: TV Munson and Son; 1909.
76. Bailey LH: The species of grapes peculiar to North America. *Gentes Herbarum* 1934, 3:151-244.
77. This P, Lacomb T, Thomas MR: Historical origins and genetic diversity of wine grapes. *Trends Genet* 2006, 22:511-519.
78. McGovern PE: *Ancient wine. The search for the origins of viticulture*. Princeton: Princeton University Press; 2003.
79. Moore MO:*Vitaceae*. [*Treatment from the Jepson manual*] [http://ucjeps.berkeley.edu/interchange/I_treat_indexes.html .
80. De Queiroz K: Different species problems and their resolution. *BioEssays* 2005, 27:1263-1269.
81. Nixon KC, Wheeler QD: An amplification of the phylogenetic species concept. *Cladistics* 1990, 6:211-223.
82. Comeaux BL: Nomenclatural clarification of the name *Vitis simpsonii* Munson (Vitaceae). *SIDA* 1987, 12:279-286.
83. *Genetic Resources Information Network*. [http://www.ars-grin.gov/npgs/acc/acc_queries.html]; [http://www.ars-grin.gov/cgi-bin/npgs/html/tax_search.pl?Vitis
84. Doyle JJ, Doyle JL: DNA isolation from leaves. *Phytochemical Bulletin* 1987, 19:11-15.
85. Lodhi MA, Ye GN, Weeden NF, Reisch BI: A simple and efficient method for DNA extraction from grapevine cultivars and *Vitis* species. *Plant Mol Biol Rep* 1994, 12:6-13.
86. Wheeler DL, Barrett T, Benson DA, Bryant SH, Canese K, Chetvernin V, Church DM, DiCuccio M, Edgar R, Federhen S, Geer LY, Kapustin Y, Khovayko O, Landsman D, Lipman DJ, Madden TL, Maglott DR, Ostell J, Miller V, Pruitt KD, Schuler GD, Sequeira E, Sherry ST, Sirotkin K, Souvorov A, Starchenko G, Tatusov RL, Tatusov TA, Wagner L, Yaschenko E: Database resources of the National Center for Biotechnology Information. *Nucl Acids Res* 2006, 36(suppl 1):D13-D21.
87. Labate JA, Baldo AM: Tomato SNP discovery by EST mining and resequencing. *Mol Breed* 2005, 16:343-349.
88. Rozen S, Skaletsky HJ: *Primer3: A software component for picking PCR primers.* [https://sourceforge.net/project/showfiles.

php?group_id=112461]; public WWW server at [http://Fokker.wi.mit.edu/primer3/input.htm]

89. Altschul SF, Gish W, Miller W, Myers EW, Lipman DJ: Basic local alignment search tool. *J Mol Biol* 1990, 215:403-410
90. Filatov DA, Proseq: A software for preparation and evolutionary analysis of DNA sequence data sets. *Mol Ecol Notes* 2002, 2:621-624.
91. Cornish-Bowden A: Nomenclature for incompletely specified bases in nucleic acid sequences: recommendations. *Nucleic Acids Res* 1985, 13:3021-3030.
92. Chenna R, Sugawara H, Koike T, Lopez R, Gibson TJ, Higgins DG, Thompson JD: Multiple sequence alignment with the Clustal series of programs. *Nucl Acids Res* 2003, 31:3497-3500
93. Baldo AM, Francis DM, Caramante M, Robertson LD, Labate JA: AlleleCoder: a PERL script for coding co-dominant polymorphism data for PCA. *Plant genetic resources: characterization and utilization* 2011, 9:528-530.
94. Posada D, Crandal KA: The effect of recombination on the accuracy of phylogeny estimation. *J Mol Evol* 2002, 54:396-402.
95. Drummond AJ, Rambaut A: BEAST: Bayesian evolutionary analysis by sampling trees. *BMC Evol Biol* 2007, 7:214-222.
96. Kluge AG: A concern for evidence and a phylogentic hypothesis of relationships among Epicrates (Bovidae, Serpentes). *Syst Zool* 1989, 38:7-25.
97. Simmons MP, Ochoterena H: Gaps as characters in sequencebased phylogenetic analyses. *Syst Biol* 2000, 49:369-381.
98. Müller K: SeqState - primer design and sequence statistics for phylogenetic DNA data sets. *Appl Bioinf* 2005, 4:65-69.
99. Simmons M, Müller K, Norton A: The relative performance of indel-coding methods in simulations. *Mol Phylogenet Evol* 2007, 44:724-740.
100. Egan AN, Crandall KA: Incorporating gaps as phylogenetic characters across eight DNA regions: ramifications for North American Psoraleeae (Leguminosae). *Mol Phylogenet Evol* 2008, 46:532-546.
101. Wiens JJ: Combining data sets with different phylogenetic histories. *Syst Biol* 1998, 47:568-581.

102. Huson DH: SplitsTree: analyzing and visualizing evolutionary data. *Bioinformatics* 1998, 14:68-73.

103. Huson DH, Bryant DF: Application of phylogenetic networks in evolutionary studies. *Mol Biol Evol* 2006, 23:254-276.

104. Gradstein FM, Ogg JG, Smith AG, Agterberg FP, Bleeker W, Cooper RA, Davydov V, Gibbard P, Hinnov LA, House MR, Lourens L, Luterbacher HP, McArthur J, Melchin MJ, Robb LJ, Shergold J, Villeneuve M, Wardlaw BR, Ali J, Brinkhuis H, Hilgen FJ, Hooker J, Howarth RJ, Knoll AH, Laskar J, Monechi S, Plumb KA, Powell J, Raffi I, RÖhli U, *et al.*: *A geologic time scale 2004*. Cambridge: Cambridge University Press; 2005.

105. International Stratigraphic Chart http://www.stratigraphy.org/index.php/ics-chart-timescale

106. Drummond AJ, Marc A, Suchard MA, Dong Xie D, Rambaut A: Bayesian phylogenetics with BEAUti and the BEAST1.7.

107. Figtree. http://tree.bio.ed.ac.uk/software/figtree/

108. Drummond AJ, Ho SYW, Phillips MJ, Rambaut A: Relaxed phylogenetics and dating with confidence.*PLoSBiology* 2006, 4:e88.

109. HIV sequence database. [http://www.hiv.lanl.gov/content/sequence/findmodel/findmodel.html

110. Bruno WJ, Socci ND, Halpern AL: Weighted neighbor joining: a likelihood-based approach to distance-based phylogeny reconstruction.*Mol Biol Evol* 2000, 17:189-197.

111. Yang Z: PAML 4: a program package for phylogenetic analysis by maximum likelihood. *Mol Biol Evol* 2007, 24:1586-1591.

112. Posada D, Crandall KA: Modeltest: testing the model of DNA substitution. *Bioinformatics* 1998, 14:817-818.

113. Donoghue MJ: A critique of the biological species concept and recommendations for a phylogenetic alternative. *Bryologist* 1985, 88:172-181.

114. Cracraft J: Species concepts and speciation analysis. *Curr Ornithol* 1983, 1:159-187.

115. Stamatakis A: RAxML-VI-HPC: maximum likelihood-based phylogenetic analyses with thousands of taxa and mixed models. *Bioinformatics* 2006, 22:2688-2690.

116. Stamatakis A, Hoover P, Rougemont J: A fast bootstrapping algorithm for the RAxML web-servers. *Syst Biol* 2008, 57:58-771
117. Huson DH, Richter DC, Rausch C, Dezulian T, Franz M, Rupp R: Dendroscope: An interactive viewer for large phylogenetic trees. *BMC Bioinforma* 2007, 8:460.
118. Huelsenbeck JP, Ronquist F: MRBAYES: Bayesian inference of phylogeny. *Bioinformatics* 2001, 17:754-755.
119. Rambaut A, Drummond AJ: *Tracer V171*. [http://beast.bio.ed.ac.uk/Main_Page
120. Goloboff PA, Farris JS, Nixon KC: TNT, a free program for phylogenetic analysis. *Cladistics* 2008, 24:774-786.
121. Fitch WM: Toward defining the course of evolution: minimum change for a specified tree topology. *Syst Zool* 1971, 20:406-416.
122. Giribet G: TNT: tree analysis using new technology. *Syst Biol* 2005, 54:176 178.
123. Nixon KC: The parsimony ratchet, a new method for rapid parsimony analysis. *Cladistics* 1999, 15:407-414.
124. Goloboff PA: Analyzing large data sets in reasonable times: solutions for composite optima *Cladistics* 1999, 15:415-428.
125. Ronquist F: Ancestral areas and parsimony. *Syst Biol* 1994, 43:267-274.
126. Li H-L: Floristic relationships between Eastern Asia and Eastern North America. *Trans Am Philos Soc* 1952, 42:371-429.
127. Thorne RF: Major disjunctions in the geographic ranges of seed plants. *Q Rev Biol* 1972, 47:365-411.
128. Boufford DE, Spongberg SA: Eastern Asian- Eastern North American phytogeographical relationships – a history from the time of Linnaeus to the twentieth century. *Ann Missouri Bot Gard* 1983, 70:423-439.
129. Tiffney BH: Perspectives on the origin of the floristic similarity between Eastern Asia and Eastern North America. *J Arnold Arbor Harv Univ* 1985, 66:73-94.
130. Fennel JL: Inheritance studies with the tropical grape. *J Hered* 1948, 39:54 66.

131. Comeau BL: A new *Vitis* (Vitaceae) from Vera Cruz, Mexico. *Sida* 1987, 12:273-277.
132. Wan Y, Schwaninger HR, Li D, Simon CJ, Wang Y, He P: The eco-geographic distribution of wild grape Germplasm in China. *Vitis* 2008, 47:77-80.
133. Hsu J: Late Cretaceous and Cenozoic vegetation in China, emphasizing their connections with North America. *Ann Missouri Bot Gard* 1983, 70:490-508.
134. Milne RI, Abbott RJ: The origin and evolution of Tertiary relict floras. *Adv Bot Res* 2002, 38:281-314.

Chapter 3

Species Favourability Shift in Europe due to Climate Change: A Case Study for Fagus sylvatica L. and Picea abies (L.) Karst. Based on an Ensemble of Climate Models

Wolfgang Falk[1] and Nils Hempelmann[2]

[1]Bavarian State Institute of Forestry (LWF), Department Soil and Climate, Hans-Carl-von-Carlowitz-Platz 1, 85354 Freising, Germany

[2]Helmholtz-Zentrum Geesthacht Zentrum für Material- und Küstenforschung GmbH, Climate Service Center (CSC), Chilehaus, Fischertwiete 1, 20095 Hamburg, Germany

ABSTRACT

Climate is the main environmental driver determining the spatial distribution of most tree species at the continental scale. We investigated the distribution change of European beech and Norway spruce due to climate change. We applied a species distribution model (SDM), driven by an ensemble of 21 regional climate models in order to study the shift of the favourability distribution of these species. SDMs were parameterized for 1971–2000, as well as 2021–2050 and 2071–2100 using the SRES scenario A1B and three physiological meaningful climate variables. Growing degree sum and precipitation sum were calculated for the growing season on a basis of daily data. Results show a general north-eastern and altitudinal shift in climatological favourability for both species, although the shift is more marked for spruce. The gain of new favourable sites in the north or in the Alps is stronger for beech compared to spruce. Uncertainty is expressed as the variance of the averaged maps and with a density function. Uncertainty in species distribution increases over time. This study demonstrates the importance of data ensembles and shows how to deal with different outcomes in order to improve impact studies by showing uncertainty of the resulting maps.

INTRODUCTION

Climate change affects tree species because trees are sessile, long lived, and comparatively slow in reacting to a changing environment, for example, trees cannot easily colonize new habitats in a strongly fragmented landscape. Because of the slow growth rates of trees to reach maturity, the typical planning horizon in the forestry sector is, for example, for spruce 60 to 80 years in the future. Hence, forestry has to deal with the challenges of a rapidly changing environment in order to avoid breakdown of stands. The same is true for city planners or municipal authorities with regard to tree species. Therefore, forest owners and related professionals need a reliable assessment about shifts of species distribution in order to make management decisions [1].

Even though genetic resources are considered as a key factor of forest ecosystems to adapt to climate change [2], one has to keep in mind that adaptation processes are slow, that niche conservatism is described for some genera [3], and that species will not survive or at least will not have a positive population growth rate if environmental conditions at a site will not meet the physiological requirement of a species.

Species distribution models (SDMs) can be used to gain ecological insights and to predict distributions across landscapes and to extrapolate in space and time [4]. SDMs relate distribution data (e.g., occurrences at known locations) with information on the environmental conditions at these sites [4]. On a continental scale, tree species distribution is mainly caused by climate variables [5] and only on a finer scale by soil or terrain. For example, Coudun et al. showed that soil nutritional factors improve distribution models of species with a high demand of alkaline cations in the soil solution like Acer campestre [6]. Additionally, tree species distribution is influenced by management which can result in over- or underdispersion of some species. Reference [7], for example, reviews major effects on beech distribution in north-eastern Europe on the one hand due to conversion of beech stands into agricultural land and later into coniferous forests since the medieval epoch and on the other hand due to planting and promotion of beech near or even outside the reputed distribution range in Prussian times. The SDM approach assumes equilibrium between species distribution and environmental conditions. Problems might arise in some areas where a species is not present due to human impact or when the postglacial migration is not completed (under-dispersion). If those areas have comparable environmental conditions as areas with presences, a model prediction for those areas is possible.

SDMs most often use presence-absence data, the model output is a probability of occurrence. This probability can be transformed into binomial data with the help of a threshold value (e.g., [8]). According to Real et al. [9], the probabilities do not only depend on the values of the predictor variables but also on the relative proportion of presences and absences. In order to overcome this effect and to describe the environmental favourability of a site for or against a species, Real et al. [9] suggest the use of a favourability function to receive results that are independent of an unbalanced proportion of presences and absences. Especially beech has more absences than presences in the data base; therefore the favourability transformation is used in this study. Norway spruce (Picea abies (L.) Karst.) and European beech (Fagus sylvatica L.), two of the most important tree species of the humid zone of Europe referring to forestry, are examined in this study. Spruce is considered to be strongly affected by the climate change, whereas beech is considered to be a possible suitable alternative as sites become dryer and warmer in central Europe [1]. Both species are relatively unspecific to soil and terrain conditions [10, 11]; therefore, their distribution is mainly determined by climatic constraints. Thus, our study concentrates on the relation of climate and distribution. Plant growth is limited amongst others by low temperatures (e.g., [12]) and, therefore, precipitation sum (PsumVeg) and temperature sum (TsumVeg) are calculated in respect of the start and end of the growing season. Beside these two variables, minimum temperature of the dormancy (TminDorm) is used to describe species distribution.

Projected changes in temperature and precipitation patterns differ within seasons in Europe [13,14]. Furthermore, the annual cycle changes, that is, the vegetation period is enlarging [15, 16] and has to be represented by chosen climate indices (CI). Furthermore, plant growth is limited amongst others by low temperatures (e.g., [12]) and therefore physiological meaningful indices need to be related to a period of higher temperatures. If bud burst is earlier and leaf falls later in the year, the indices forcing the SDM have to take the start and end date of the vegetation period into account. Thus, forcing values are related to length of vegetation period in this study. The terms vegetation period and growing season are used synonymously and refer to a climatological definition.

When using climate data for SDM approaches, the question of appropriate climate data or models has to be solved. There are a variety of data for present conditions or future climate projections [17–19], calculated with different models, bounding conditions, and assumptions. Many studies on species distribution refer to the WorldClim database [20, 21], because these worldwide data have a very high spatial resolution [17]. WorldClim data are interpolated climate surfaces for global land areas. Future data are calculated with a statistical downscaling of the output of global circulation models (GCM) and these anomalies are applied to the WorldClim data. In general, it is hard to judge which data set is the best [22]. Therefore, the use of an ensemble of data seems to be the state-of-the-art solution to the question of the right data set. This is especially true for future climate data because there are many different data experiments with global circulation models (GCM) which in turn are forcing different dynamic regional circulation models (RCM). The GCMs are set up under defined assumptions, the so-called scenarios (Special Report on Emissions Scenarios, SRES). In this study, only the SRES A1B scenario [23] and one SDM technique has been considered in order to concentrate on differences of circulation models and downscaling methods and to show their effects on SDM output. The A1 storyline and family describe a future world of very rapid economic growth, global population that peaks in mid-century and declines thereafter, and the rapid introduction of new and more efficient technologies [23]. The technological emphasis of A1B is a balance across all sources. That means that similar improvement rates apply to all energy supply and end use technologies [23]. Cumulative emissions of greenhouse gases are in a broad range and in the middle of the different scenarios and families [23]. Due to this, A1B is often used in climate impact studies (e.g., [24]).

The framework from assumptions of boundary conditions for climate change projection to final merged results of spatial favourability distribution for tree species is given by a chain of separate steps:

$$SRES \rightarrow GCM \rightarrow RCM \rightarrow CI \rightarrow SDM \quad (1)$$

For every chain member, several different models, approaches, or assumptions can be taken into account and enlarge the number of ensemble members. Due to the usage of data and model ensembles, a corresponding number of SDM results had to be merged. This final step can be taken as an additional chain member.

In this study, we focus on the methodological framework due to the chain of steps and present an approach focusing on ensembles of different GCMs and RCMs. Investigation area is Europe, distribution models are built by means of climate data from the period 1971–2000, and maps are calculated for this period and two future timesteps (2021–2050 and 2071–2100). The study is focusing on the following issues:

- shift of climatic favourability for European beech and Norway spruce in Europe;
- incorporation of ensembles of gridded and daily resolved climate model variables into a SDM framework;
- accentuation of uncertainties of the final results that derive from the use of an ensemble of climate models.

DATA AND DATA PREPROCESSING

The presented SDM approach is driven by downscaled regional climate model data. The data are taken from the ENSEMBLE-Project [13, 19] where data experiments with different RCMs, forced by different GCMs, are available. Within the ENSEMBLES-Project 22, data experiments have been realized with regional climate over Europe, all following the assumptions of the IPCC emission scenario A1B [23]. The experiments cover the time period from 1951 to 2100. Seven datasets only completed up to the year 2050; thus only 15 of the 22 realizations are available for the far future. The combination of RCM and appropriate forcing GCM is shown in Table 1. For all ensemble members (single datasets), daily resolved values of near surface air temperature (at 2 m) and surface precipitation were taken. As indicated in the table, we excluded ensemble members, because of implausible values at least in some European regions (e.g., very high precipitation over Spain) which led to physiological implausible SDMs. In case of spruce we left out one realization and in case of beech two (marked in Table 1). The decision to exclude these datasets is based on the shape of the single response curves.

Table 1: Ensemble members for distribution calculation and the appropriate GCMs and RCMs. Detailed description of the models is reported in [13]

GCM	ARPEGE				BCM		CGCM3		ECHAM5		
RCM	CNRM-RM4.5	CNRM-RM5.1	DMI-HIRHAM5	DMI-HIRHAM5	METNOHIRHAM	SMHI-RCA	OURANOSMRCC4.2.1	C4IRCA3	DMI-HIRHAM5	ICTP-REGCM3	KNMI-RACMO2
1971-2000	x	x	x	x	x	x	x*	x	x	x	x
2021-2050	x	x	x	x	x	x	x*	x	x	x	x
2071-2100	–	x[1]	x	x	–	x	–	–	x[1]	x	x

GCM	ECHAM5				HadCM3Q0			HadCM3Q3		HadCM3Q16	IPSL
RCM	MPI-REMO	SMHI-RCA	ETHZ-CLM	METO-HC-HadRM3Q0	METNOHIRHAM	UCLM-PROMES	VMGO-RRCM	METO-HC-HadCM3Q3	SMHI-RCA	METO-HC-HadCM3Q16	MPI-REMO
1971-2000	x	x	x	x	x	x	x**	x	x	x	x
2021-2050	x	x	x	x	x	x	x**	x	x	x	x
2071-2100	x	x	x	x	–	–	–	x	x	x	x

X: realized SDM; [1]year 2100 deleted; *skipped for beech; **skipped for beech and spruce.

X: realized SDM; [1]year 2100 deleted; *skipped for beech; **skipped for beech and spruce.

Downloaded ENSEMBLE data were preprocessed in order to receive a uniform file format. To ensure a later on merging, all datasets were remapped to a unique domain from 15.125°W to 39.875°E and 35.125°N to 70.125°N and a grid of 0.25° resolution (according to the E-OBS data grid [25]). Beside these remappings, all units were harmonised for timesteps (in UNIX epoch, seconds after 01.01.1970), daily surface temperature (in Kelvin) and precipitation (in mm/day). All preprocessing of regional climate data were realized with Climate Data Operators (CDO) version 1.5.3 [26]. All in all, we used 21 climate datasets and 3 time slices (see Table 1). Due to too many missing data in nine data experiments for the year 2100, this year was excluded completely for the appropriate realizations.

Beside data describing the climate conditions, observation data of the geographical distribution of present occurrence of tree species is used in this study. The data for beech and spruce were taken from the ICP Forests large-scale forest condition monitoring (Level I) from the period 1987 to 2007. In order to receive a somewhat regular grid, some plots were left out where the density of plots was considerably more than one plot per 225 km^2 (e.g., due to shift of inventory plots within a country or duplicates). This led to more than 7500 inventory plots in a resolution of approximately one plot per 225 km^2 across Europe [27].

The available information for crown condition within the monitoring data were transformed into presence and absence data as required by the SDM approach of this study. Inventory data often comprise fallacious absences [28, 29], for example, due to impact by forest management for long time periods. In case of beech there is fallacious absence in southern Scandinavia because there beech is not surveyed within the monitoring program. This means that the data include absences where a species could live successfully but it is not present due to other reasons than unfavourable climate conditions. The term fallacious absence used by [28] refers to this theoretical concept and does not mean that data are improper in a technical way. As we aim to describe the climatic favourability, we came up against fallacious absences by using a vegetation map [30] that provides an overview of vegetation units and tree species composition. We attached the information about vegetation units to the Level I plots and transformed it into presence and absence data for beech and spruce. Finally, the two columns were merged. Therefore, a zero means no presence in both of the datasets, whereas a one means a presence in at least one of the datasets. In order to receive a clear signal of the altitudinal alpine distribution limit, we added 56 pseudoabsences above the alpine tree line. The resulting final data set consisted of 7596 plots covering most of Europe with a lower spatial coverage in the eastern European parts in the chosen spatial extent.

We aimed to find independent data for validation and therefore merged data from three different sources: first, the left out Level I data; second, Level II (ICP Forests Intensive Monitoring Programme, http://icp-forests.net/page/level-ii/) data, and, third, data from the European Forest Genetic Resources Programme (http://www.euforgen.org/, resp., http://www.eufgis.org/). All data were transformed into presence and absence data, no treatment for false absences was carried out and the use of the data from the Genetic Resources Programme was restricted to stands with more than one species in order to increase the probability that a full species inventory was behind the data. All in all, a maximum of 2135 stands could be used for validation. The data were unevenly distributed but covered most parts of Europe. Presences and absences of spruce fitted better to rough distribution maps (http://www2.biologie.uni-halle.de/bot/ag_chorologie/areale/index.php?sprache=E/) than presences and absences of beech. Validation data were combined with climate datasets. Because of missing values, for example, close to the shore line in some combinations of GCM with RCM, we left out all those validation sites that did not contain the complete climate information of all 21 ensemble members.

METHODS

The overall method is to relate species occurrence data to an ensemble of climate data and to transfer this relation to climate projections for this ensemble. The distribution modeling, quality assessment, and merging of the final maps is described in this section (Figure 1). The section refers to the last to chain links of the chain mentioned in the introduction.

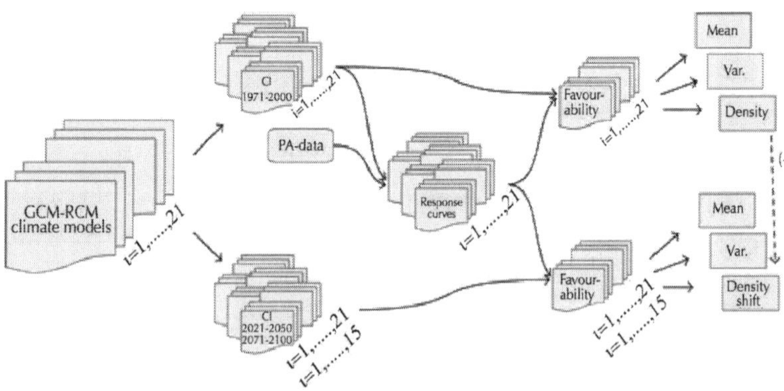

Figure 1: General framework of the separate methodical steps to model species distribution based on an ensemble of climate models with 21 respective 15 members.

The SDMs are forced by climate indices which refer to a vegetation period. These indices were calculated for each grid box. A grid box is the smallest spatial unit of the climate datasets which have a resolution of 25 km. As the vegetation period is the base for the indices, the start and end day in the year of this period has to be determined first. The vegetation period, or synonymously the thermal growing season, in this study is defined as a period with daily mean temperatures ≥5°C for spruce and ≥10°C for beech. We used a running average over 11 days for the calculation of the start and end day using the daily resolved temperature data. Temperatures had to be above or below the threshold for at least five consecutive days. Furthermore, some limitations based on a day length criterion were set as default: the vegetation period was not allowed to start before first of March (60 yday) and could not end before first of July (152 yday) but had to end latest after first of November (305 yday).

If five consecutive days with temperature above, respectively, below 5° (10°)C were detected, the first day of the consecutive timesteps is taken as the start respective end day in the year. The following equations are describing the dedications of vegetations periods begin as well as their ending:

$$V_s = \text{yday if } \{\text{Trun}_{(\text{yday}+0,\dots\dots,4)} \geq T_{\text{base}} \quad \text{yday} \geq 60\} \text{ else} 60 \qquad (2)$$

$$V_e = \text{yday if } \{\text{Trun}_{(\text{yday}+0,\dots\dots,4)} \leq T_{\text{base}'} \quad 152 \leq \text{yday} \leq 305\} \text{ else } 305, \qquad (3)$$

With

yday = day in the year

V_s= start of vegetation period (yday)

V_e= end of vegetation period (yday)

T run = running mean with 11 timesteps of daily mean temperature.

The length of the vegetation period (VP) can be defined as

$$VP = V_e - V_s. \qquad (4)$$

The following indices are used, which are derived from daily resolved temperature and precipitation values considering the information on the length of the vegetation period.

Precipitation sum in the vegetation period: PsumVeg [mm]

$$\text{psum veg} = \sum_{yday=V_s}^{V_e} p_{yday'} \quad (5)$$

growing degree days in the vegetation period: TsumVeg [°C]

$$\text{Tsum Veg} = \sum_{yday=V_s}^{V_e} (T_{yday} - T_{base}), \quad (6)$$

total minimum temperature outside the vegetation period: TminDorm [°C]

$$\text{T min Dorm} = \min(\text{Trun}_{(V_s \geq yday \geq Ve)}), \quad (7)$$

With

p= daily precipitation amount (mm)

T= daily mean temperature (°C).

The selection of physiological meaningful variables was done according to our hypotheses on the climate dominated distribution of beech and spruce on the continental scale and according to preliminary work (variable importance of a boosted regression tree analysis; see [4] for details on boosted regression trees).

TminDorm is detected based on a smoothed temperature line to reduce misdetection of extreme values and stored for every year. The smooth is again the running average over 11 days. Finally, a mean over the respective 30 year time period was calculated for each index and for each grid box. The climate indices were calculated for each ensemble member and time period separately and joined with the species distribution data.

In order to understand the time dependent trends of the climate data, time series of the indices were calculated as a mean over a small test region for each climate data set. We chose the Bavarian Forest as a test region 12.625° to 13.875°E and 48.875°N to 49.125°N as this landscape is suitable for growing beech and spruce.

We used generalized additive models (GAMs, [31–34]) using the R-package MGVC [33] (R-project version 2.15.0 (2012-03-30)) to calibrate the single SDMs. GAMs are semiparametric extensions of generalized linear models. They allow fitting of response curves with a nonparametric smoothing function instead of parametric terms.

This improves the exploration of species responses to environmental gradients. GAMs were used without interaction terms in order to build parsimonious models. An individual GAM was fitted to each climate ensemble member of the present day time slice (1971–2000) for beech and spruce. Response curves were calculated for the climate indices (TsumVeg, PsumVeg and TminDorm) and checked for physiological plausibility according to [35]. We used thin plate regression splines [34]; the degree of smoothness in the package mgcv is in general estimated by a generalized cross validation (GCV) criterion within certain limits set, for example, by the dimension *k* of the basis used to represent the smooth term. –k-values were first set to 3 and adapted (maximum k=6) in some cases in order to meet assumptions on physiological plausibility. These values affect the maximum possible degree of freedom for each term of the GAM. One climate data set was excluded due to implausible precipitation data (no decrease of favourability at low precipitation for beech and spruce in Europe, that is, no distribution limit due to drought) and another one for beech (implausible response curve of variable growing degree days) after checking the single response curves for plausibility. All in all, a separate calibration of the GAM for each data set was done.

Different quality criteria were taken from the GAM output and additional parameters like the area under the receiver operating curve (AUC), sensitivity, and specificity were calculated for learning and validation data with the R package PresenceAbsence [36]. Additionally, a tenfold cross validation was run, leaving out one-tenth of the data each time and using these data for validation. The GAMs were used for the calculation of distribution maps. These maps depict the probability of occurrence for Picea abies and Fagus sylvatica for each ensemble member and time slice (present day and future). In a further step, probabilities were transformed into favourabilities according to [9] in order to allow an easy comparison between the species. This is due to the fact that this transformation corrects the effect of the frequency of presences in the learning data (prevalence) on the probabilities [9]. The output is a floating number between 0 (no) and 1 (maximum favourability) and describes the environmental favourability of a site for a species.

The last step was the merging of the separate results for each ensemble member to a final result. Two strategies were realized: an average over all appropriate ensemble results as well as a pseudodensity calculation. The average favourability (mean) of a grid box was only calculated if grid boxes contain valid values for all ensemble members. If a grid box contains a "Not a Number" (NaN) value, this 25 km rectangle was removed from the averaging procedure (e.g., at coastlines). Additionally, the variance was calculated. The second way of integrating the different results was the calculation of a density map. Calculating density functions is a concept of merging ensembles with a large number of members. In this study we received 15 (far future), respectively, 21 (near future) members and therefore called the results due to the low number "pseudodensity function." Favourability output was converted into a binomial response (site is favourable or not) using a threshold of 0.5 (the prevalence transformed to the scale of favourability) according to [8]. For each ensemble result, the favourability of ≥0.5 and ≤1 was taken as a one and otherwise as a zero. The pseudodensity is the proportion of ones at each grid cell. If all ensemble members have a one, then the ratio or pseudodensity is one; if half of the ensemble members have a one and the other half a zero, then the result is 0.5. The distribution shift was determined by subtracting the pseudodensity of the future predictions from that of the present time.

RESULTS

Due to the chain from future scenario assumption to species distribution projection, several interim data were calculated. The following section presents results and figures that illustrate differences in the climate data ensembles and these differences—as a consequence—can be found again in the distribution models. Single response curves of the GAM approach try to illustrate this as well as averaged distribution maps of beech and spruce. The section ends with the two different averaging methods that are explained in Section 3 and shows the respective distribution shift for future time slices.

Climate Indices Ensemble

The output of the climate model runs is afflicted with high uncertainties, but in this work the focus is on the handling of different climate model output and these uncertainties are not taken into account. Therefore, uncertainties of the final distribution maps within the ensemble approach first of all are due to differences in the climate input data. For example, precipitation data of the 21 ensemble members do have different maximum amounts of precipitation. They agree in high precipitation sums in Alpine regions but differ markedly in the amount of precipitation at the coastlines (not shown). Since the climate data differ, the derived indices differ as well. As an example, Figure 2 shows the length of the vegetation period calculated with a baseline of 10°C for a test region in Eastern Bavaria. While the ensemble mean shows a vegetation period of around 130–150 days in the present time slice, single ensemble members differ with a range of more than 80 days. This variation is decreasing in the future due to the reduction of datasets as only 15 datasets were available for the far future. Additionally, in the case of the 5°C-baseline the fixed limit of 244 days (vegetation period only between April and October) is reached in some years and reduces variation. As climate indices (precipitation sum and growing degree days) base on the length of the vegetation period, these indices inherit this variability. During the century, an extension of the vegetation period of 15 days is described by the climate data within the first half and of up to 35 days by the end of the century. Generally, growing season is longer for spruce as an evergreen conifer than for deciduous beech due to the different baselines (5°C, resp., 10°C) which results in an earlier start and later end day of vegetation period for spruce.

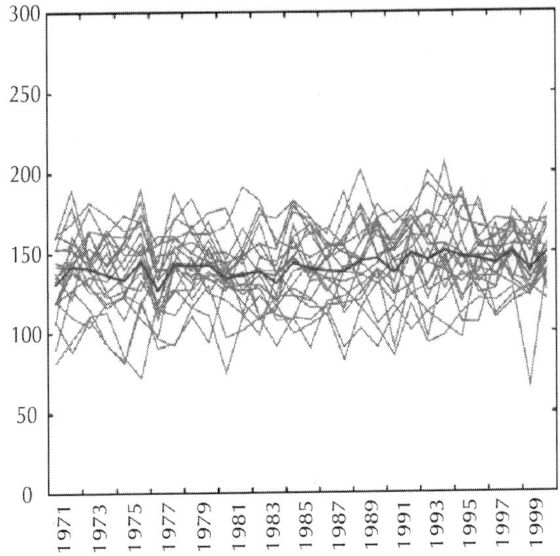

(a)Time slice 1971–2000

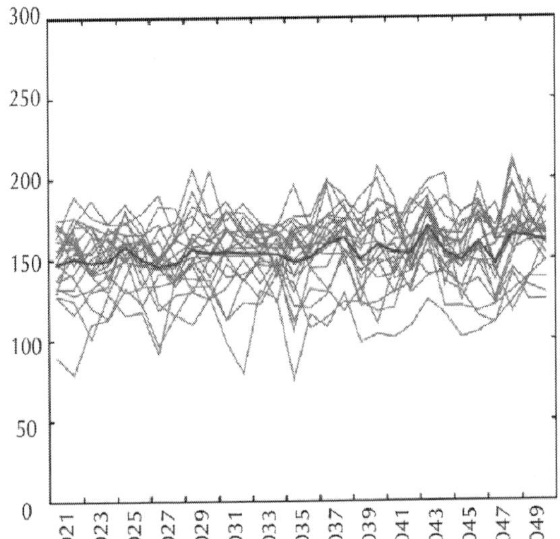

(b) Time slice 2021–2050

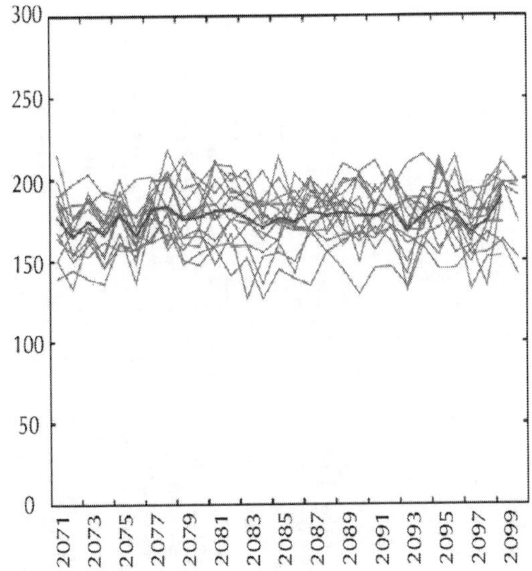

(c) Time slice 2071–2100

Figure 2: Time series of length of vegetation period with the baseline of 10°C for the test region Bavarian Forest for all ensemble members (grey lines). Red line shows the mean.

Figure 3 shows a summary of the growing degree days calculated for spruce at the inventory plots with the data of the time slice 1971–2000. This sample illustrates the differences of the climate data ensemble members which in turn have consequences for species distribution modeling. The differences are even stronger for precipitation sum (not shown) because precipitation is harder to model than temperature.

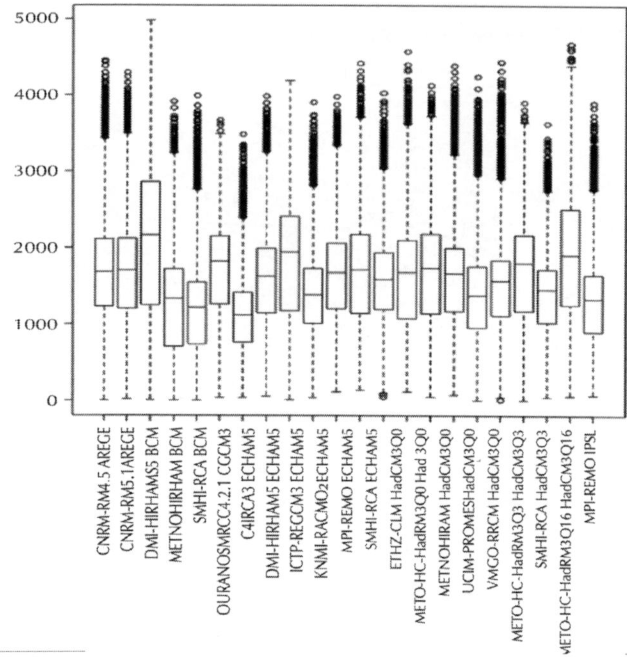

Figure 3: Comparison of the climate index growing degree days (°C) at the inventory plots. The ensemble of climate data is shown for the period 1971–2000. Growing degree days are calculated depending on the length of the vegetation period defined as a period above 5°C. Base line of the temperature sum is 5°C (input data for modeling the distribution of Picea abies).

GAM Ensemble

The resulting single response curves of the GAMs of spruce are shown in Figure 4. Even though there is a general accordance of most models in the shape of the response curves, in some cases there are marked deviations due to the different climate data (e.g., Figure 3). The differences are pronounced where the data are scarce, for example, in the range of high precipitation sums which occur at high altitudes or at some coastlines. Additionally, response curves diverge due to different maxima, for example, of precipitation sum, and therefore there is a shift in the range of values. The marked differences of the input data lead to marked shifts of the response curves but the shape which describes the type of response seems to accord for many

SDMs. Basically, GAMs agree in describing a distribution limit at mild winter temperature (3–5°C minDorm), high temperature sums in the vegetation period (>2000–2500°C TsumVeg), and low precipitations sums (<200 mm PsumVeg). The individual GAMs differ in describing the relationship of low temperature sums and high precipitation sums with the occurrences of spruce. In case of beech (not shown), response curves of TminDorm are bell-shaped, TsumVeg describes a distribution limit at high temperatures, and PsumVeg shows a distribution limit due to drought.

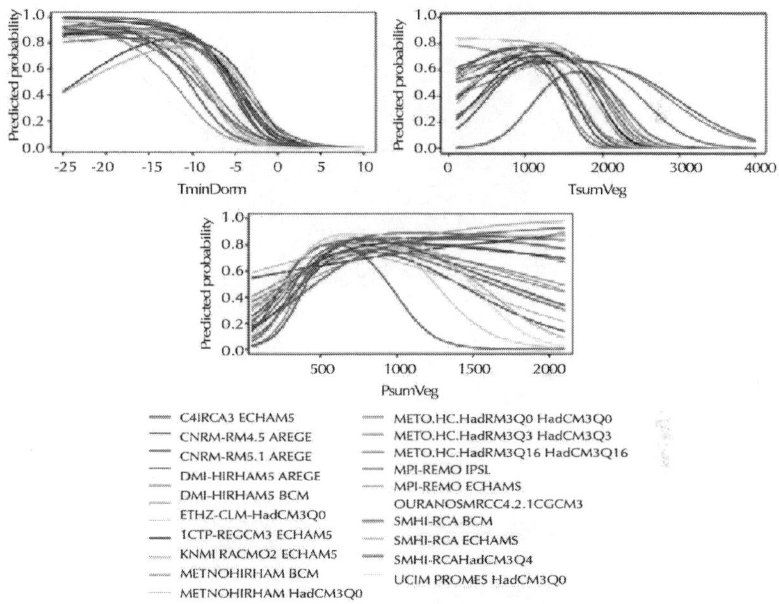

Figure 4: Single response curves of the GAMs for all climate data and speciesPicea abies. Single response curves are calculated for each climate index (TminDorm, TsumVeg, and PsumVeg) while keeping the two others constant at the mean value. Single response curves give a hint on model behavior and physiological plausibility. Legend names are combinations of RCM and GCM.

Favourability Ensemble

The 30 years averages of the climate indices are the basis of the SDMs that are used to calculate the spatial distribution of spruce and beech in terms of favourability. Accordingly, the resulting maps represent averaged climate conditions for the three periods 1971–2000, 2021–2050, and 2071–2100. As we cannot present all results for each ensemble member, we show the final result as a union of the entire ensemble by averaging the single members of each time period (Figures 5and 6). Additionally, we present the corresponding variability in the same figures.

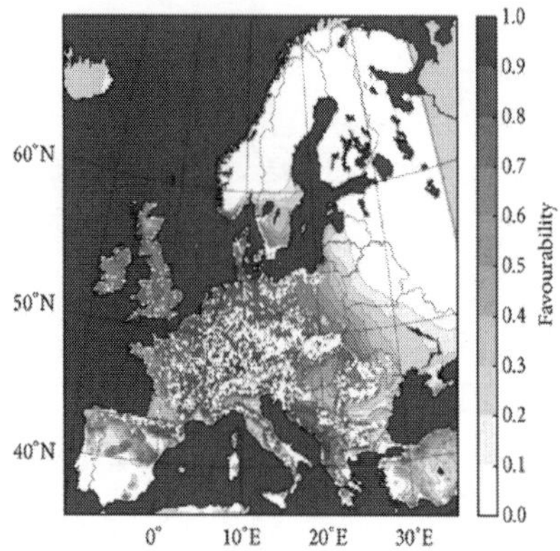

(a)

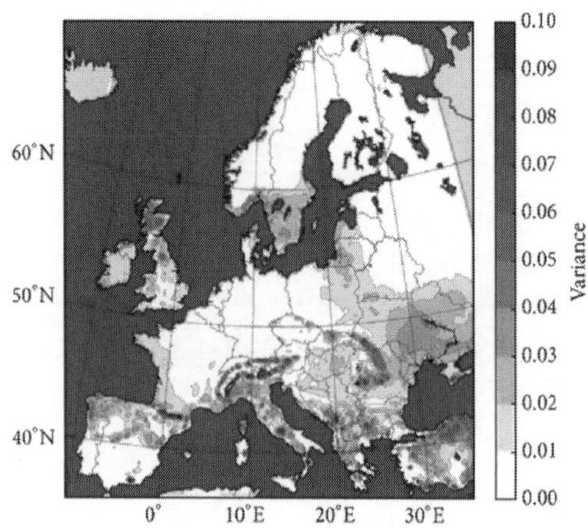

(b)

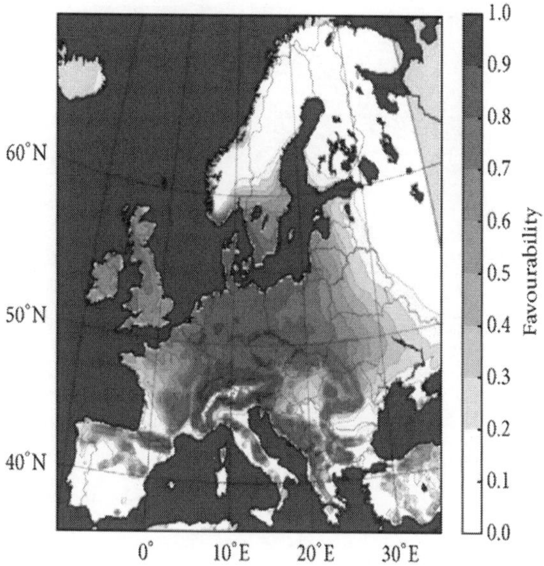

60°N
50°N
40°N
0°
10°E
20°E
30°E
1.0
0.9
0.8
0.7
0.6
0.5
0.4
0.3
0.2
0.1
0.0
Favourability

(c)

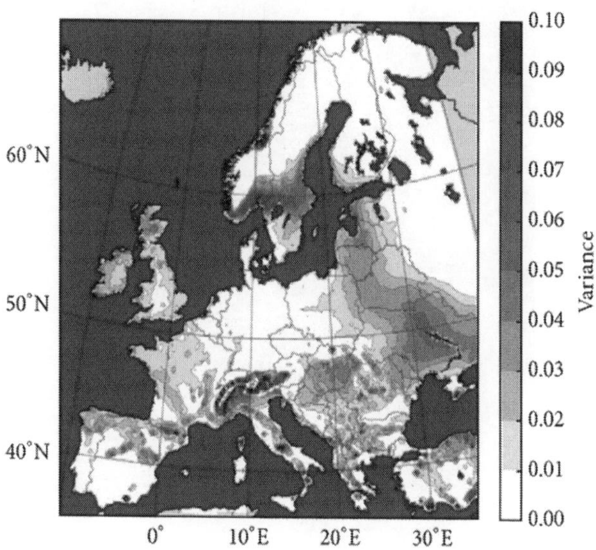

60°N
50°N
40°N
0°
10°E
20°E
30°E
0.10
0.09
0.08
0.07
0.06
0.05
0.04
0.03
0.02
0.01
0.00
Variance

(d)

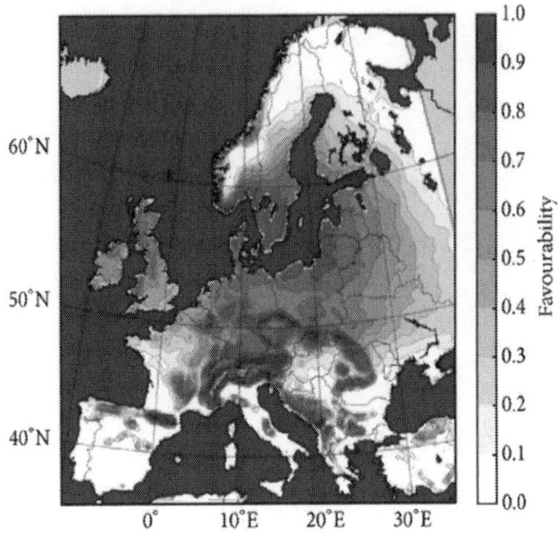

(e)

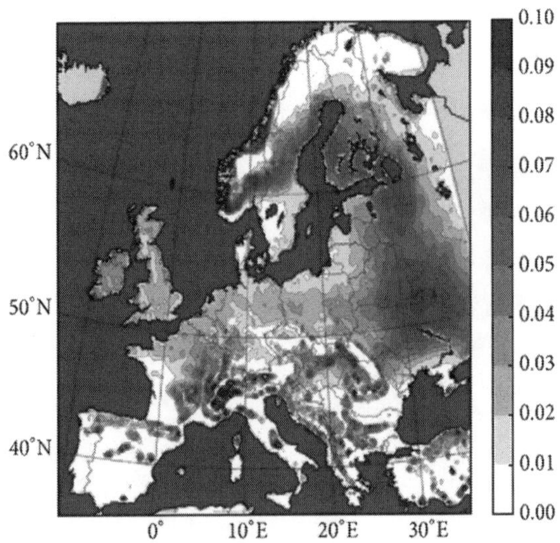

(f)

Figure 5: Mean ((a), (c), and (e)) and variance ((b), (d), and (f)) of the spatial distribution of Fagus sylvatica (favourability according to [9]). White dots in (a) are beech occurrences in the learning data Mean and variance of (a), (b), (c), and (d) underlie an ensemble of 20 climate data, whereas the ensemble of (e) and (f) consist only of 15 members.

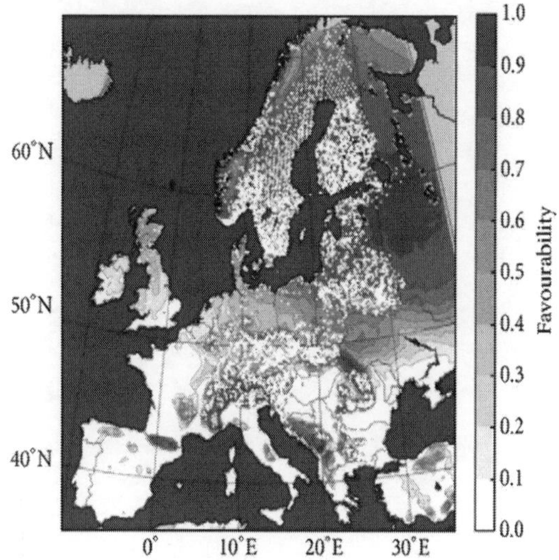

(a)

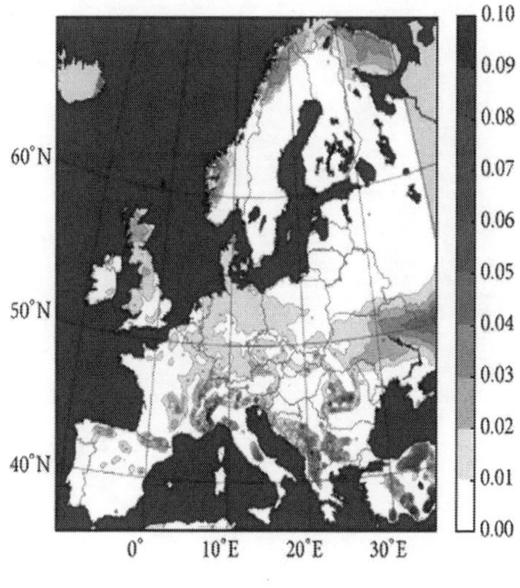

(b)

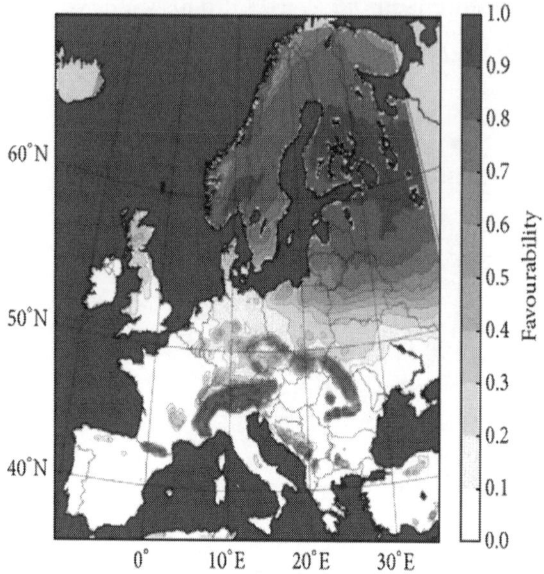
60°N
50°N
40°N
0°
10°E
20°E
30°E
1.0
0.9
0.8
0.7
0.6
0.5
0.4
0.3
0.2
0.1
0.0
Favourability

(c)

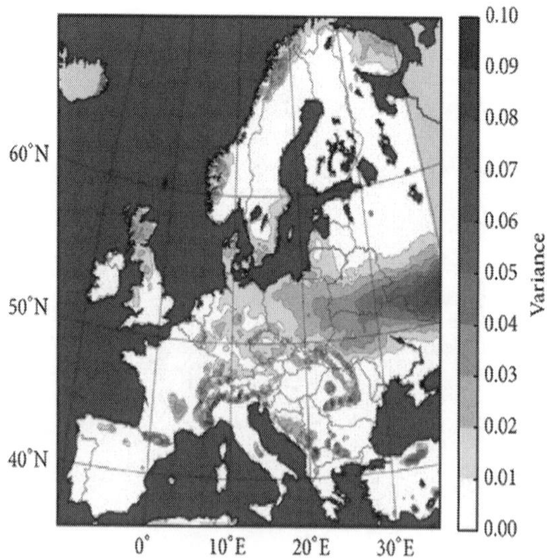
60°N
50°N
40°N
0°
10°E
20°E
30°E
0.10
0.09
0.08
0.07
0.06
0.05
0.04
0.03
0.02
0.01
0.00
Variance

(d)

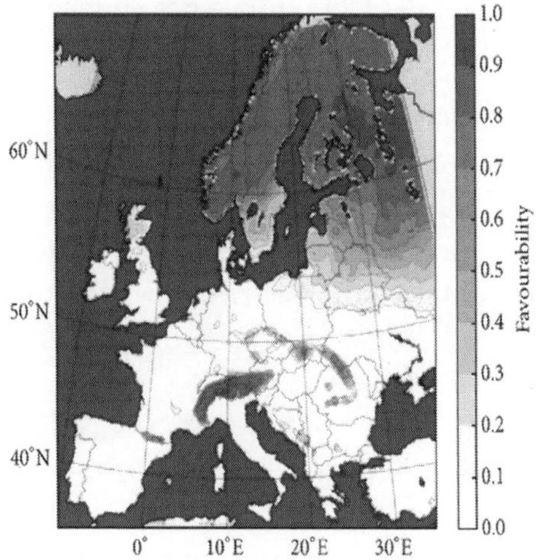

(e)

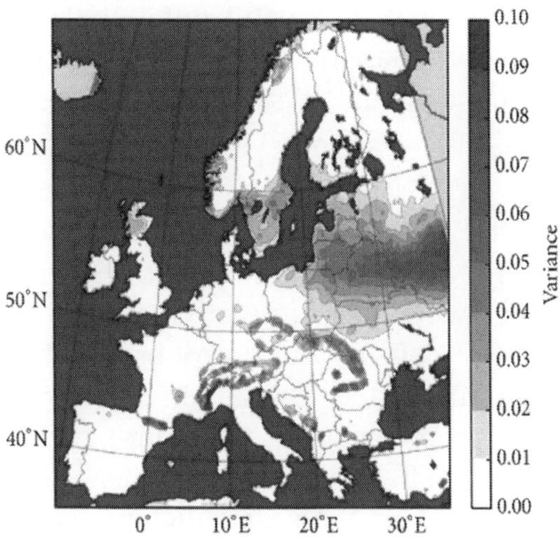

(f)

Figure 6: Mean ((a), (c), and (e)) and variance ((b), (d), and (f)) of the spatial distribution of Picea abies (favourability according to [9]). White dots in (a) are spruce occurrences in the learning data. Mean and variance of (a), (b), (c), and (d) underlie an ensemble of 21 climate data, whereas the ensemble of (e) and (f) consist only of 15 members.

The favourability map of beech shows a high accordance with the presences of the training data (white dots in the figure). This is underlined by high quality measures, for example, a mean AUC (area under the receiver operating curve; see [37]) of 0.90 with a minimum of 0.87 for the learning data. Using the prevalence of 0.32 as a threshold, the SDMs on average classify 90 percent of presences and 76 percent of absences correct which leads to 80 percent of the training data correctly classified. The Kappa value (see, e.g., [37]) is determined on average as 0.59 with a minimum of 0.53. Quality measures of the 10-fold cross-validation are in the same order of magnitude with the same mean and 2 to 9 per cent lower minimum values. For the validation data, the figures decrease: a mean AUC of 0.76, 85 percent of presences but only 53 percent of absences are on average correctly classified. The mean Kappa value is 0.35. Besides the present day distribution and quality of favourability maps, the figure characterizes the possible impact of climate change on the species distribution. Beech seems to be stable in middle Europe (e.g., Germany) until 2050 with only moderate shifts from west to north-east but the shift is anticipated to increase by the end of the century (calculation with 15 ensemble members). The changes mostly affect areas with low present day favourabilities of beech like western France.

Overall, the highest variability between the maps for beech is at the edges of distribution and it increases in the future especially in regions with a rising favourability in eastern and north-eastern Europe. Hotspots of variability can be found around the Alps, the Pyrenees as well as certain parts of the Carpathian arc, in the east Ukraine, and in a belt from St. Petersburg, Russia, around the gulf of Bothnia (northernmost arm of the Baltic Sea), via Sweden up to the Norwegian coastline. Generally, regions with a predicted favourability of 0.3 to 0.6 often also show a high variance of more than 0.06. The variance of the present day ensemble is low and thus points to a reliable result.

The situation for spruce is different (Figure 6). First of all, the distribution edge, for example, in Northern Germany or France appears to be not matched as good as in the case of beech (white dots with low favourability). These areas are highly affected by a strong northward shift of spruce's favourability by the end of the century. Second, the predicted distribution shift for spruce is larger than for beech and all resulting maps agree. That means that there is a low variance of the mean. Highest variances are found at the edge of the investigation area (a hotspot in Ukraine stretching into Belarus and Russia marking the southern limit of the distribution). Models have a mean AUC of 0.91 (minimum 0.89), a Kappa value of 0.67 (minimum 0.63), classify 83 percent correctly, a sensitivity of 0.89, and a specificity of 0.77 calculated with the training data and the prevalence as a threshold value (0.51). Again, 10-fold cross-validation results in the same order of magnitude with 1 to 6 per cent lower minimum values. Validation with the independent data provides the following quality measures: a mean AUC of 0.88, a mean Kappa value of 0.61, and 81 percent correctly classified.

Pseudodensity

Figure 7 presents the calculated pseudodensity functions and the differences between present day and future distribution of beech and spruce. There is a high accordance of the models describing the distribution expressed in a narrow transition belt from high density values to low ones and only a few differing GAM outcomes at the distribution edges (e.g., southeast Europe in case of beech). A few areas with low density in Italy or southwest France are backed as suitable by presences (white dots in Figure 7(a)). Interpreted as a measure of accordance or disaccordance, the pseudodensity helps to illustrate areas where the results are reliable due to high accordance (e.g., areas where 80 percent of the SDM agree). Additionally, the figures show the distribution shift due to climate change in terms of density shifts, allowing a clear picture of where favourable areas for growing beech and spruce decline (values below zero, red areas) and where new areas are gained (values above zero, green areas). Certainly, there is a high accordance between averages and pseudodensity maps.

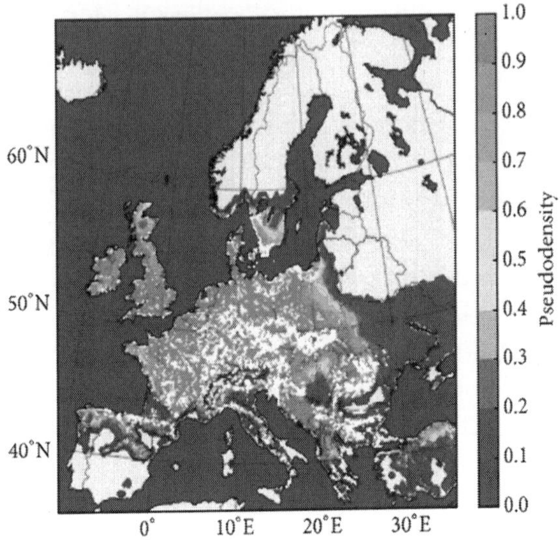
1.0
0.9
0.8
0.7
0.6
0.5
0.4
0.3
0.2
0.1
0.0
Pseudodensity
60°N
50°N
40°N
0°
10°E
20°E
30°E

(a)

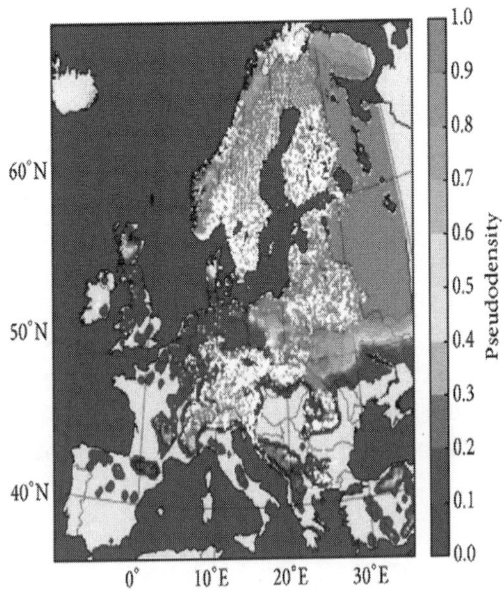
1.0
0.9
0.8
0.7
0.6
0.5
0.4
0.3
0.2
0.1
0.0
Pseudodensity
60°N
50°N
40°N
0°
10°E
20°E
30°E

(b)

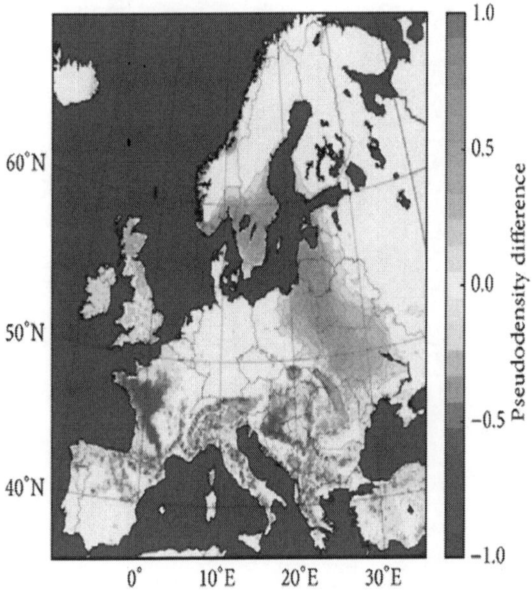
60°N
50°N
40°N
0°
10°E
20°E
30°E
1.0
0.5
0.0
−0.5
−1.0
Pseudodensity difference

(c)

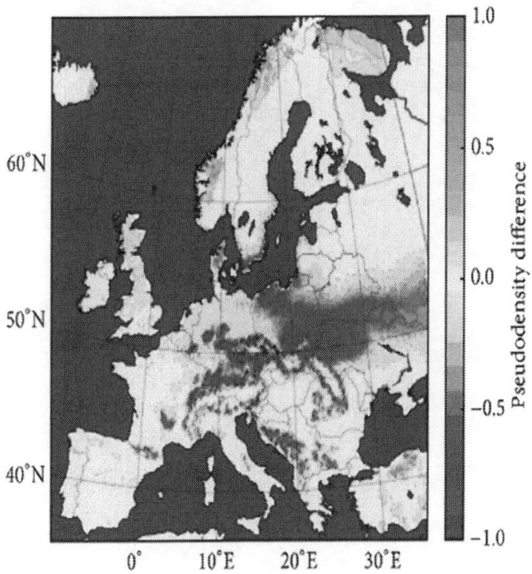
60°N
50°N
40°N
0°
10°E
20°E
30°E
1.0
0.5
0.0
−0.5
−1.0
Pseudodensity difference

(d)

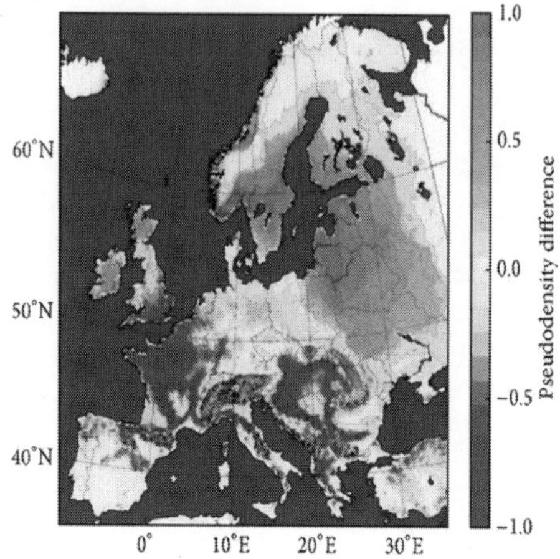

(e)

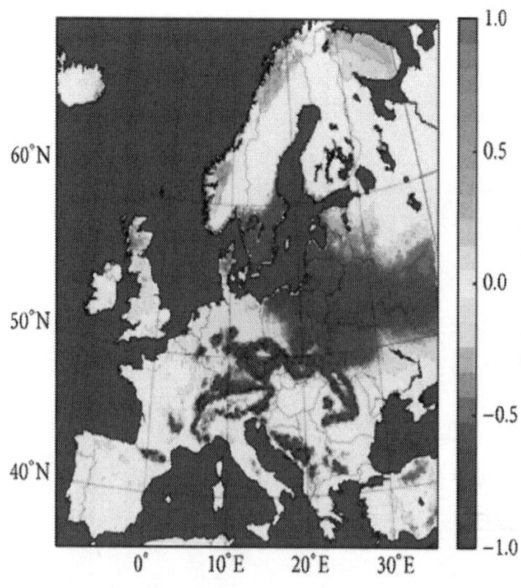

(f)

Figure 7: Shift of distribution of Fagus sylvatica ((a), (c), and (e)) and Picea abies ((b), (d), and (f)) expressed as difference of density distribution of the SDM ensemble in reference to time slice 1971–2000. White dots in (a) and

(b) are beech respective spruce occurrences in the learning data. Density distribution in (c) and (d) underlies an ensemble of 20 respective 21 members, whereas the ensemble of (e) and (f) consists of only 15 members.

The results for beech as shown in Figures 7(a), 7(c), and 7(e) indicate a high accordance of different maps, in the middle of the century a slow north-eastern distribution shift that accelerates by the end of the century. In 2021–2050 beech stands at the western (western France) and southern distribution limit (Apennine Mountains in Italy, Hungary and some parts of southeastern Europe) will be affected first. These areas are widening by the end of the century also affecting the south of England, Ireland, and northern and northeastern Germany. On the other hand, new favourable regions for growing beech are gained in higher altitudes and northeastern Europe. In the case of spruce (Figures 7(b), 7(d), and 7(f)), the main spatial distribution in terms of pseudodensity function is situated around the Baltic Sea. The distribution includes the whole Scandinavian Peninsula except the higher altitudes of the Norwegian Mountains. The southern distribution border is characterized by a relatively large transition zone of lower model agreement along 50°N for the more continental regions and shifting northwards within an oceanic influence. Further south, the agreement of the entire ensemble is large again in the higher altitudes of the Carpathian Arc, the Swiss Alps, and in a few regions of the Pyrenees and the Balkan regions. Beside smaller parts of the Massif Central in France, the higher altitudes of the German low mountain ranges are also pointed out as suitable sites for spruce under present climatic conditions described with the SDMs. Nevertheless, the presences (white dots in Figure 7(b)) in northern Germany, the Benelux countries, and in France emphasizes the problems of many spruce stands in warmer climates today. They are in climatically unfavourable areas outside their natural distribution range and further warming as projected by scenarios indicates a high risk of decline. Sites with a raising favourability for spruce in the future can be found in the Scandinavian mountains and at the actual northern distribution edge. The distribution shift also shows that favourability for spruce declines in Germany (lower mountain ranges and alpine upland), Czech Republic, northeast Poland, Pyrenees and French Massif Central, and southern Sweden.

DISCUSSION

This section provides a discussion of the major steps of the presented modeling chain including the climate data, the indices driving the SDMs, the SDM approach, and the resulting final maps.

Climate Data and Ensemble Modeling

Using projection ensembles in SDMs is a common technique in order to reduce variance and uncertainty and to raise the accuracy of species distribution models [38–40]. Whereas many studies focus on ensembles of SDM model classes, our focus is on the climate data, an approach becoming more important in the SDM literature [41]. Perhaps due to limited access to climate data or problems in handling huge data volumes, ensembles often are limited to a couple of combinations of GCM and RCM (e.g., [1], but see [24]). Beside the assumptions made on the development of greenhouse gas emissions, the realizations of climate data via GCM and RCM differ depending on the particular GCM and RCM used. This introduces an additional range of uncertainty into the model chain. Regionalized downscaled climate data differ markedly, even when using only one emission scenario (Figures 2 and 3). We therefore concentrate in this study only on the different realizations of GCMs and RCMs and not on different emission scenarios.

A state-of-the-art ensemble model set is that of the ENSEMBLES Project. Lorenz and Jacob [42] compared the RCMs driven with quasiobserved lateral boundary conditions with gridded observational data [25] for near surface temperatures and concluded an underestimation of the past temperature trends as well as a wide range of the results of the RCMs. Jacob et al. [14] reported an uncertainty of several degrees as well as uncertainties in precipitation among the projected model data. The uncertainties are increasing during the century up to Δ 2.5 K for running averages and Δ 30% in case of precipitation. The changes vary during the seasons and different regions. The difficulty of judging the single members of the ensemble leads to an ensemble without weighting their members [22, 43].

Not all ensembles members have a realization over the whole century; in seven cases the future projections were only resized to the year 2050. Therefore, a smaller variance at the end of the century does not mean a lower uncertainty.

Climate Indices: Use of Daily Climate Data

We focused on three climate indices describing the environmental conditions driving species distribution. An alternative for using temperature and precipitation is the use of aridity indices (e.g., [44]). They have the advantage of combining two variables and therefore taking interactions into account. The disadvantage is that they are normally developed for climate data with a monthly or yearly resolution and that they are therefore weaker related to plant physiology. Two of the indices used in this work—temperature and precipitation in the growing season, PsumVeg and TsumVeg—depend on the length of this period, which can be estimated with daily climate data. Daily resolved data are suitable for the tackled questions but require methods of handling large data amounts.

As shown in Figure 2, the vegetation period is slightly enlarging with rising temperatures in the future projections which is in accordance with extrapolations made from phenological observations [16, 45, 46]. The projected temperature rise for the 21st century has a twofold effect on the calculated indices. On the one hand, there are higher temperatures leading to a higher temperature sum. On the other hand, the reference base for the calculation of the temperature and precipitation sum—the length of the vegetation period—is enlarging too. This can intensify a trend in case of the rising temperature sums as well as compensate decreasing precipitation sum values due to more days which are taken into account to the end of the century.

We use the vegetation period for two of the selected indices because precipitation and temperature during the physiological active phase are most important for water supply and evapotranspiration and therefore tree growth [12] and thus for species distribution (e.g., see [47], a study on drought and mortality of spruce). Due to the large climatic differences within Europe, a fixed time period presumably has a weaker relation with tree growth. We use a simple but straightforward approach to define the growing season (mean temperature threshold and day length criterion). The distribution of beech and spruce are located in climate regions with thermal seasons. The detection defined by five consecutive days higher respective lower than a threshold based on a smoothed temperature line is a very simple but useful approach. There are more sophisticated ones (e.g., [45, 46, 48]), but due to the variances within the ensemble members there is no benefit expected with a more complex approach for vegetation period detection.

The chosen simple definition of the vegetation period may not be suitable for all European regions, that is, in southern Europe where the boundaries 1st of March and 31st of October are reached too often, due to their default exclusion. This happens especially in the case of spruce for which we assumed a temperature limit of 5°C for considerable transpiration and other physiological growth processes (Raspe pers. com., [45, 48–50]). We think our approach still does make sense because the distribution models are developed for species of the temperate to boreal zone and regions with high temperatures almost all over the year are far beyond the distribution range of spruce and beech. An important point is that we opened the door for using regionalized daily climate data. In future works more sophisticated indices concentrating on drought events in the growing season or drought indices can be implemented.

Species Distribution Data and Modeling

Use of Vegetation Map Data for the Treatment of Fallacious Absence

Species distribution models are only as good as the calibration data. Hence, if environmental data do not explain the distribution well because of unaccounted effects like forest management practice (e.g., preference of one species over another due to cultivation traditions or economic reasons), postglacial migration borders due to competition, or land use effects, models may have high prediction errors. As forest inventory data are affected by human impact, we cannot assume an equilibrium between species traits and distribution leading to errors due to fallacious absences [28, 29]. That means that a species is absent although the site is favourable. The use of expert knowledge is one possibility to treat these fallacious absences (see [44]). So on the one hand, the used vegetation map helps to improve the distribution data.

On the other hand, it might introduce some uncertainty due to its scale. In our opinion, the advantages outweigh the disadvantages because the distribution maps now give a general picture where spruce and beech potentially can grow. Whether the data describe the full potential of a species is not clear and it is obvious that our models are influenced by the human impact on species distributions. In case of spruce, we assume that plantations show the distribution limit in continental Europe. Ireland might be an example where our data is not complete and do not properly represent the potential distribution: models describe low to medium favourability for Ireland despite presences in the learning data. Absences dominates over the few presences in this area. The maps show the potential distribution area [51]. This information is valuable for forest management decisions dealing with adaptation to climate change. Whether species will colonize this potential distribution depends on many factors like, for example, migration speed and barriers, dispersal and establishment limitations, and pests that, for example, can be described with mechanistic models [52].

The independent validation data were not corrected for fallacious absences and therefore reflect the realized distribution and show the differences between potential and realized distribution. Spruce is a tree species that is planted outside its natural distribution range with the help of forest management. Therefore, there are not so much fallacious absences and realized distribution is almost equal to potential distribution. Hence, validation results are in the same magnitude as cross-validation results (AUC, Kappa, percent correctly classified stands). The situation is different in case of beech, where quality measures of the validation with independent data are lower compared to cross-validation results. Beech is missing at sites where vegetation experts judge it to be within its natural distribution range (fallacious absences). It might be missing due to forest management and preference of spruce or oaks, for example, for economic reasons. Fallacious absences lead to higher error rate (percent correctly classified absences, Kappa value) when comparing these absences with a model that describes the potential distribution. That means that there could be a gap between actually realized distribution and potential distribution as depicted by vegetation experts because the cross-validation as the second validation method backs the model quality. Independent validation data could have been corrected for fallacious absences but then there would not be obtained more information of this method than by the cross-validation procedure.

Use of Climate Data as Proxies for Water Supply

We used climate data to describe the distribution of spruce and beech. Soil influences species distribution, too (e.g., preference of sites with high base saturation in case of Acer campestre [6]). However, spruce and beech are highly competitive, widely distributed in Europe, and do not have a special preference to soil conditions (e.g., nutrients) or certain relief positions [10, 11]. Soil is also part of the hydrological cycle and therefore does matter when drought events are the reason for distribution limits. Generally, it is difficult to find adequate European soil data and the influence of soil physics on water availability on a European scale is small compared to the effects of precipitation and temperature. On a local scale it is much more important to incorporate soil data into SDMs. In case of the Level I data, the European demonstration project BioSoil might fill the gap of missing soil data in the future [53]. Nevertheless, climate data as proxies for water supply in the growing season appear to be sufficient to reproduce the general distribution pattern of most tree species in Europe and to risk a glimpse into what climate change might mean for the distribution of these species. The picture will change and distribution models improve when taking into account frequency, intensity, and duration of drought [54]. Daily climate model data prepare the technical ground to focus on extreme events.

Species Distribution Modeling Technique—GAM

Following the argumentation of Wang et al. [41] and referring to the work of Marmion et al. [39], we concentrate on one SDM technique. Reference [39] recommends either a single best method or a consensus approach of different methods like realized within the Biomod framework [40]. Other than [41], we do not use a machine-learning technique like Random Forest or a Boosted Regression Tree technique despite their good performance [4, 52, 55], but use GAM, a well-established semiparametric regression technique [55]. GAM can be handled to model smoothed continuous gradients meeting the continuum theory [56]. Additionally, it can be run in a parsimonious way using only a few dose variables, whereas the strength of data driven machine learning techniques amongst others is to deal with a high number of variables and their interactions [4]. It is especially important to avoid overfitting when balancing between model fit and predictive performance because overfitting may reduce transferability [4, 57]. We calibrated the GAMs in order to give a generalized picture of how the distribution of spruce and beech is determined by precipitation and temperature. Therefore, GAMs were run with low –K values (basis dimension of the penalized regression smoother, K-1 is the upper limit of the degrees of freedom and influences computational efficiency) which leads to a high degree of smoothness and therefore results in very general trends (see partial response curves in Figure 4). Thus, we avoid overfitting and the SDMs are as reliable as possible in terms of transferability or more precise forecasting [38, 44].

Calibration of GAMs and Plausibility of Individual Climate Data Experiments

One step in the presented species distribution modeling was the individual tuning of the single response curves in order to receive physiological plausible models according to some fundamental hypotheses about the relationship between climate and species distribution. In doing this, we follow [35]. Elith et al. [58] also suggested to deliberately control the fit of models, for example, by using exploratory techniques like checking response curves for their contribution to the final outcome. They even went one step further and successfully integrated information from mechanistic models in order to enhance the reliability of correlative predictions. When calibrating the GAMs individually for each climate data set, we also tried to reach a high sensitivity (correctly modeled presences) because this is a requirement for describing the ecological potential of a species. Nevertheless, response curves differ which is a result of the different climate data and highlights the uncertainty of models in some areas, for example, at the edges of distribution where presences are scarce. As we cannot judge the quality of the different datasets, we use the ensemble technique in order to obtain a more complete picture of the distribution and a possible change with rising temperatures by the end of the century. A comparison of differences in input data (Figure 3) and SDMs (Figure 4) reveals the dependence of the SDMs on the climate data and can be used to cautiously judge the quality of input data for species distribution modeling: if there is no drought limit for the distribution of beech and spruce within a European data base, then this data set seems to be inappropriate, at least for a species distribution modeling approach. Overall, lower levels of temperature or precipitation leads to leftwards shift of single response curves in Figure 4. Thus, the SDMs do not differ markedly but the climate indices.

Transformation of Probabilities into Favourabilities

We transformed the probability of occurrence as provided by a GAM into favourabilities according to [9] in order to compare the two species with different presence/absence ratios (prevalence) easily. According to [9], "favourability values can be regarded as a degree of membership of the fuzzy set of sites whose environmental conditions are favourable to the species." As we want to describe the potential distribution area of spruce and beech, favourabilities seem to be an adequate quantity to show where spruce and beech can grow today and in the future.

Effect of Climate Change on Species Distribution in Europe

On a European scale it is obvious that spruce as a mountainous species is more affected by climate change than beech. This means that there are many spruce stands that will have a high risk of mortality due to warmer and drier stand conditions. There are different ways of managing risk in managed forests like shortening of rotation period, thinning [59], and in the long term transformation into stable mixed stands. We work on the species level not taking provenances into account due to the European scale and the fact that information on provenances are scarce. It can be expected that on a stand level, changes in vitality will be even more dramatic and adaptation to drier conditions will not keep pace with the projected warming. Therefore, our attempt treating the species as a whole and leaving genetic differences out might lead to an optimistic picture for single stands and a pessimistic one for long term adaptation processes. The latter is because we model the whole distribution range of the species and do not take a closer look on the distribution edges. With the help of forest management, today spruce is already grown beyond its natural distribution range at relatively warm sites. This means that stands that will fall outside the range in the future might survive for a certain time but the mortality risk is increasing strongly. Mortality might not be caused directly by drought but drought leads to lower vitality and subsequent pest attack. It is clearly visible that beech can be an alternative species on stands that will be too dry for spruce in the future because drought is a major discriminating factor at the warm and dry distribution limit between favourabilities for beech and spruce. Beech can withstand drier conditions because of deeper rooting and the chance to reduce leaf area in extreme situations in summer by a premature fall of leaves. Nevertheless, the distribution of beech has a drought limit too. With lower competition (e.g., growth) due to drought [60] beech fails to shade out other species.

Final Maps: Mean and Density

Using ensemble modeling techniques entails handling of the multiple output generated. A simple but straightforward approach is to calculate a mean and the variance if there is no evidence of differences in quality of the resulting models or maps [22]. Otherwise, a weighted mean (or median) should be favored [22]. We present this approach (simple mean) as it leads to a map describing the whole range between suitable and unsuitable environments without losing information due to classification. Typically, species distribution models that are built on binomial distribution data lead to binomial output maps with the help of a threshold value. This leads to a clear picture of the final maps but has the disadvantage of losing information. A compromise is the use of multiple thresholds [44], linking the two approaches. We present another way of handling the ensemble output using one threshold, the prevalence [8], and thus combining binomial maps. we call it a "pseudodensity function" as the amount of ensemble members is limited to 20, respectively, 21, or 15, in case of the far future. The maps of the two approaches (mean and pseudodensity function) look similar to a certain degree but the density map has sharper distribution limits due to its binomial input maps.

CONCLUSION

Projections of the translocation of climatic favourability for beech and spruce can be done with SDMs forced by climate model and species presence/absence data because the growth of these tree species is mainly depending on climate conditions. Thus, the presented approach can be used for tree species favourability projections where edaphic factors can be neglected on a European scale.

The study shows consequences that can be expected for spruce and beech due to climate change with regard to their distribution in Europe. The general trend is a north-eastward and altitudinal shift of climatological favourability for both species, though stronger in case of spruce. These outcomes are in accordance with other studies [1, 24] and can be used as a basis for risk assessment. The altitudinal shift is only relevant for beech because spruce already is part of the Alpine tree line and the area above this line with sufficient soil substrate for tree growth is limited. In general, the distribution shift will strongly affect the middle European spruce stands and underlines the need for stand conversion of pure spruce stands into mixed forests. Beech can be part of mixed stands and therefore used to stabilize spruce stands but certain beech stands at the warm and dry edge of distribution might need a conversion too. The presented work gives an example on how gridded daily resolved values of climate parameters, like temperature at 2 m and surface precipitation, can be used to describe the climatic site conditions for tree species. With these daily resolved climate model data very specific indices can be calculated. We present a rather simple approach to determine the vegetation period just to give an idea of the potential of temporal highly resolved data experiments.

The use of an ensemble with 21 members reduces the risk of a wrong projection. The study shows the importance and advantages of ensemble projections by pointing out regional differences of the uncertainties for each time slice. The outcome of an impact model like a SDM with climate data as input depends strongly on these differences. Beside a wide spread and common use of an arithmetic average combined with the appropriate variance, pseudodensity functions are another tool to visualize uncertainty originating from the climatic data ensemble. The individual check for physiological plausibility reduces bias and uncertainty deriving from the SDM framework, though this source of uncertainty cannot be eliminated completely although the used GAM technique, separately trained for each ensemble member, minimizes the individual bias of the different climate model members within the ensemble.

OUTLOOK

The data handling with daily resolved climate model data and the connection to species distribution modeling is done with the presented work. Now that the frame is set, extension into different ways can be made: the focus can be set on variability and extreme events [54] and their influence on species distribution. For example, drought indices or deviations from long term mean might provide a better explanation of species distribution than simple mean values [54]. Certainly, the models could be run for other species, or other SDM techniques.

The Coordinated Regional Downscaling Research Experiment (CORDEX http://wcrp-cordex.ipsl.jussieu.fr/) is the successor of ENSEMBLE-Project and is forced by the RCP scenarios [61] instead of those of the SRES scenario families used in the ENSEMBLE-Project. If these data are available, an ensemble with more members, higher spatial and temporal resolution, and new emission scenarios can be taken to force impact models like SDMs. Amongst other things, the CORDEX data will reduce the data preprocessing due to observing some standards. Future work will be done with these new climate data.

To reduce the workload or invite impact modelers with limited access to climate data, the shown GAM approach will be available in a virtual computing environment (C3Grid-INAD) and therefore can help to process large datasets under the control of the user (e.g., use other climate indices for sensitivity or extreme events studies or different definitions of the vegetation period).

ACKNOWLEDGMENTS

The ENSEMBLES data used in this work was funded by the EU FP6 Integrated Project ENSEMBLES (Contract no. 505539) whose support is gratefully acknowledged. The tree species data are derived from the first grid (Level I) data. These data are surveyed under the International Co-operative Programme on Assessment and Monitoring of Air Pollution Effects on Forests operating under the UNECE Convention on Long-Range Transboundary Air Pollution (ICP Forests, http://icp-forests.net/) and provided by the Programme Co-ordinating Centre (PCC) of ICP Forests in Hamburg, Germany, whose support is gratefully acknowledged.

REFERENCES

1. M. Hanewinkel, S. Hummel, and D. A. Cullmann, "Modelling and economic evaluation of forest biome shifts under climate change in Southwest Germany," Forest Ecology and Management, vol. 259, no. 4, pp. 710–719, 2010.
2. S. Kapeller, M. J. Lexer, T. Geburek, J. Hiebl, and S. Schueler, "Intraspecific variation in climate response of Norway spruce in the eastern Alpine range: selecting appropriate provenances for future climate," Forest Ecology and Management, vol. 271, pp. 46–57, 2012.
3. B. Huntley, P. J. Bartlein, and I. C. Prentice, "Climatic control of the distribution and abundance of beech (Fagus L.) in Europe and North America," Journal of Biogeography, vol. 16, no. 6, pp. 551–560, 1989.
4. J. Elith, J. R. Leathwick, and T. Hastie, "A working guide to boosted regression trees,"Journal of Animal Ecology, vol. 77, no. 4, pp. 802–813, 2008.
5. B. Huntley, P. M. Berry, W. Cramer, and A. P. McDonald, "Modelling present and potential future ranges of some European higher plants using climate response surfaces,"Journal of Biogeography, vol. 22, no. 6, pp. 967–1001, 1995.
6. C. Coudun, J.-C. Gégout, C. Piedallu, and J.-C. Rameau, "Soil nutritional factors improve models of plant species distribution:

an illustration with Acer campestre (L.) in France,"Journal of Biogeography, vol. 33, no. 10, pp. 1750–1763, 2006.

7. A. Bolte, T. Czajkowski, and T. Kompa, "The north-eastern distribution range of European beech—a review," Forestry, vol. 80, no. 4, pp. 413–429, 2007.
8. C. Liu, P. M. Berry, T. P. Dawson, and R. G. Pearson, "Selecting thresholds of occurrence in the prediction of species distributions," Ecography, vol. 28, no. 3, pp. 385–393, 2005.
9. R. Real, A. M. Barbosa, and J. M. Vargas, "Obtaining environmental favourability functions from logistic regression," Environmental and Ecological Statistics, vol. 13, no. 2, pp. 237–245, 2006.
10. G. von Wühlisch, "EUFORGEN technical guidelines for genetic conservation and use for European beech (Fagus sylvatica)," Tech. Rep., Bioversity International, Rome, Italy, 2008.
11. T. Skrøppa, "EURFORGEN technical guidelines for genetic conservation and use for Norway spruce (Picea abies)," Tech. Rep., International Plant Genetic Resources Institute, Rome, Italy, 2003.
12. F. Went, "The effect of temperature on plant growth," Annual Review of Plant Physiology and Plant Molecular Biology, vol. 4, pp. 347–362, 1953.
13. P. van der Linden and J. Mitchell, "ENSEMBLES: climate change and its impacts: summary of research and results from the ENSEMBLES project," Tech. Rep., Met Office Hadley Centre, London, UK, 2009.
14. D. Jacob, K. Bulow, L. Kotova, C. Moseley, J. Perterson, and D. Rechid, "CSC—Report 6: Regionale Klimaprojektionen für Europa und Deutschland: Ensembles-Simulationen für die Klimafolgenforschung," Tech. Rep., Climate Service Center (CSC), 2012.
15. A. Menzel, T. H. Sparks, N. Estrella et al., "European phenological response to climate change matches the warming pattern," Global Change Biology, vol. 12, no. 10, pp. 1969–1976, 2006.
16. N. Estrella, T. H. Sparks, and A. Menzel, "Effects of temperature, phase type and timing, location, and human density on plant phenological responses in Europe," Climate Research, vol. 39, no. 3, pp. 235–248, 2009.

17. R. J. Hijmans, S. E. Cameron, J. L. Parra, P. G. Jones, and A. Jarvis, "Very high resolution interpolated climate surfaces for global land areas," International Journal of Climatology, vol. 25, no. 15, pp. 1965–1978, 2005.
18. P. Brohan, J. J. Kennedy, I. Harris, S. F. B. Tett, and P. D. Jones, "Uncertainty estimates in regional and global observed temperature changes: a new data set from 1850," Journal of Geophysical Research, vol. 111, no. D12, Article ID D12106, 2006.
19. C. D. Hewitt, "The ensembles project: providing ensemble-based predictions of climate changes and their impacts," The Eggs Newsletter, vol. 13, pp. 22–25, 2005.
20. A. Jiménez-Valverde and J. M. Lobo, "Threshold criteria for conversion of probability of species presence to either-or presence-absence," Acta Oecologica, vol. 31, no. 3, pp. 361–369, 2007.
21. R. J. Hijmans and C. H. Graham, "The ability of climate envelope models to predict the effect of climate change on species distributions," Global Change Biology, vol. 12, no. 12, pp. 2272–2281, 2006.
22. A. P. Weigel, R. Knutti, M. A. Liniger, and C. Appenzeller, "Risks of model weighting in multimodel climate projections," Journal of Climate, vol. 23, no. 15, pp. 4175–4191, 2010.
23. IPCC, "IPCC special report emission scenarios: summary for policymakers: a special report of IPCC working group III," Tech. Rep., Intergovernmental Panel on Climate Change, 2000.
24. M. Hanewinkel, D. A. Cullmann, M. J. Schelhaas, G. J. Nabuurs, and N. E. Zimmermann, "Climate change may cause severe loss in the economic value of European forest land,"Nature Climate Change, vol. 3, pp. 203–207, 2013.
25. M. R. Haylock, N. Hofstra, A. M. G. Klein Tank, E. J. Klok, P. D. Jones, and M. New, "A European daily high-resolution gridded data set of surface temperature and precipitation for 1950–2006," Journal of Geophysical Research D, vol. 113, no. 20, Article ID D20119, 2008.
26. U. Schulzweida, L. Kornblueh, and R. Quast, Climate Data Operators. User's Guide, MPI for Meteorology Brockmann Consult, 2009.

27. M. Lorenz, "International co-operative programme on assessment and monitoring of air pollution effects on forests—ICP forests," Water, Air, and Soil Pollution, vol. 85, no. 3, pp. 1221–1226, 1995.
28. A. H. Hirzel and G. Le Lay, "Habitat suitability modelling and niche theory," Journal of Applied Ecology, vol. 45, no. 5, pp. 1372–1381, 2008.
29. P. L. Zarnetske, T. C. Edwards Jr., and G. G. Moisen, "Habitat classification modeling with incomplete data: pushing the habitat envelope," Ecological Applications, vol. 17, no. 6, pp. 1714–1726, 2007.
30. U. Bohn and R. Neuhäusl, "Interaktive CD-ROM zur Karte der natürlichen Vegetation Europas/Interactive CD-Rom Map of the Natural Vegetation of Europe.Mafistab/Scale 1:2 500 000. Teil 1: Erläuterungstext mit CD-ROM, Teil 2: Legende, Teil 3: Karten," Tech. Rep., Landwirtschaftsverlag, Münster, Germany, 2003.
31. T. Hastie and R. Tibshirani, Generalized Additive Models, Chapman and Hall, 1990.
32. S. N. Wood and N. H. Augustin, "GAMs with integrated model selection using penalized regression splines and applications to environmental modelling," Ecological Modelling, vol. 157, no. 2-3, pp. 157–177, 2002.
33. S. Wood, "Mixed GAM computation vehicle with GCV/AIC/REML smoothness estimation," 2012.
34. S. Wood, Generalized Additive Models: An Introduction With R, vol. 66, CRC Press, 2006.
35. K. H. Mellert, V. Fensterer, H. Küchenhoff et al., "Hypothesis-driven species distribution models for tree species in the Bavarian Alps," Journal of Vegetation Science, vol. 22, no. 4, pp. 635–646, 2011.
36. E. A. Freeman and G. Moisen, "PresenceAbsence: an R package for PresenceAbsence analysis," Journal of Statistical Software, vol. 23, no. 11, pp. 1–31, 2008.
37. A. H. Fielding and J. F. Bell, "A review of methods for the assessment of prediction errors in conservation presence/absence models," Environmental Conservation, vol. 24, no. 1, pp. 38–49, 1997.

38. M. B. Araújo and M. New, "Ensemble forecasting of species distributions," Trends in Ecology & Evolution, vol. 22, no. 1, pp. 42–47, 2007.
39. M. Marmion, M. Parviainen, M. Luoto, R. K. Heikkinen, and W. Thuiller, "Evaluation of consensus methods in predictive species distribution modelling," Diversity and Distributions, vol. 15, no. 1, p. 5969, 2009.
40. W. Thuiller, B. Lafourcade, R. Engler, and M. B. Araújo, "BIOMOD-a platform for ensemble forecasting of species distributions," Ecography, vol. 32, no. 3, pp. 369–373, 2009.
41. T. Wang, E. M. Campbell, G. A. O'Neill, and S. N. Aitken, "Projecting future distributions of ecosystem climate niches: uncertainties and management applications," Forest Ecology and Management, vol. 279, pp. 128–140, 2012.
42. P. Lorenz and D. Jacob, "Validation of temperature trends in the ENSEMBLES regional climate model runs driven by ERA40," Climate Research, vol. 44, no. 2-3, pp. 167–177, 2010.
43. M. Marmion, M. Luoto, R. K. Heikkinen, and W. Thuiller, "The performance of state-of-the-art modelling techniques depends on geographical distribution of species," Ecological Modelling, vol. 220, no. 24, pp. 3512–3520, 2009.
44. W. Falk and K. H. Mellert, "Species distribution models as a tool for forest management planning under climate change: risk evaluation of Abies alba in Bavaria," Journal of Vegetation Science, vol. 22, no. 4, pp. 621–634, 2011.
45. A. Menzel, Phänologie von Waldbäumen unter sich ändernden Klimabedingungen: Auswertung der Beobachtungen in den internationalen phänologischen Gärten und Möglichkeiten der Modellierung von Phänodaten [Ph.D. thesis], Technical University München, München, Germany, 1997.
46. A. Menzel and P. Fabian, "Growing season extended in Europe," Nature, vol. 397, no. 6721, p. 659, 1999.
47. S. Solberg, "Summer drought: a driver for crown condition and mortality of Norway spruce in Norway," Forest Pathology, vol. 34, no. 2, pp. 93–104, 2004.
48. K. von Wilpert, Die Jahrringstruktur von Fichten in Abhängigkeit vom Bodenwasserhaushalt auf Pseudogley und Parabraunerde.

Ein Methodenkonzept zur Erfassung standortsspezifischer Wasserstreßdisposition [Ph.D. thesis], Institut für Bodenkunde und Waldernährungslehre der Albert-Ludwigs-Universität, Freiburg im Breisgau, Germany, 1990.

49. R. Man and P. Lu, "Effects of thermal model and base temperature on estimates of thermal time to bud break in white spruce seedlings," Canadian Journal of Forest Research, vol. 40, no. 9, pp. 1815–1820, 2010.

50. M. Hannerz, "Evaluation of temperature models for predicting bud burst in Norway spruce," Canadian Journal of Forest Research, vol. 29, no. 1, pp. 9–19, 1999.

51. M. B. Araújo and A. T. Peterson, "Uses and misuses of bioclimatic envelope modeling,"Ecology, vol. 93, no. 7, pp. 1527–1539, 2012.

52. L. Iverson, A. Prasad, S. Matthews, and M. Peters, "Lessons learned while integrating habitat, dispersal, disturbance, and life-history traits into species habitat models under climate change," Ecosystems, vol. 14, pp. 1005–1020, 2011.

53. R. Hiederer, E. Micheli, and T. Durrant, Evaluation of Biosoil. Demonstration Project. Soil Data Analysis, European Union, 2011.

54. N. E. Zimmermann, N. G. Yoccoz, T. C. Edwards Jr. et al., "Climatic extremes improve predictions of spatial patterns of tree species," Proceedings of the National Academy of Sciences of the United States of America, vol. 106, no. 2, pp. 19723–19728, 2009.

55. A. Guisan, N. E. Zimmermann, J. Elith, C. H. Graham, S. Phillips, and A. T. Peterson, "What matters for predicting the occurrences of trees: techniques, data, or species› characteristics?" Ecological Monographs, vol. 77, no. 4, pp. 615–630, 2007.

56. M. Austin, "Spatial prediction of species distribution: an interface between ecological theory and statistical modelling," Ecological Modelling, vol. 157, no. 2-3, pp. 101–118, 2002.

57. C. F. Randin, T. Dirnböck, S. Dullinger, N. E. Zimmermann, M. Zappa, and A. Guisan, "Are niche-based species distribution models transferable in space?" Journal of Biogeography, vol. 33, no. 10, pp. 1527–1539, 2006.

58. J. Elith, M. Kearney, and S. Phillips, "The art of modelling range-shifting species," Methods in Ecology and Evolution, vol. 1, no. 4, pp. 330–342, 2010.

59. M. Kohler, J. Sohn, G. Nägele, and J. Bauhus, "Can drought tolerance of Norway spruce (Picea abies (L.) Karst.) be increased through thinning?" European Journal of Forest Research, vol. 129, pp. 1109–1118, 2010.

60. A. S. Jump, J. M. Hunt, and J. Penuelas, "Rapid climate change-related growth decline at the southern range edge of Fagus sylvatica," Global Change Biology, vol. 12, no. 11, pp. 2163–2174, 2006.

61. R. H. Moss, J. A. Edmonds, K. A. Hibbard et al., "The next generation of scenarios for climate change research and assessment," Nature, vol. 463, no. 7282, pp. 747–756, 2010.

Chapter 4

Higher Fine-Scale Genetic Structure in Peripheral than In Core Populations of a Long-Lived and Mixed-Mating Conifer - Eastern White Cedar (Thuja occidentalis L.)

Madhav Pandey[1,2] and Om P Rajora[1]

[1]Canadian Genomics and Conservation Genetics Institute, Faculty of Forestry and Environmental Management, University of New Brunswick, 28 Dineen Drive, Fredericton, NB E3B 5A3, Canada

[2]Current address: Department of Plant and Soil Science, Texas Tech University, 15th and Detroit Avenue, Lubbock, Texas 79409, USA

ABSTRACT

Background

Fine-scale or spatial genetic structure (SGS) is one of the key genetic characteristics of plant populations. Several evolutionary and ecological processes and population characteristics influence the level of SGS within plant populations. Higher fine-scale genetic structure may be expected in peripheral than core populations of long-lived forest trees, owing to the differences in the magnitude of operating evolutionary and ecological forces such as gene flow, genetic drift, effective population size and founder effects. We addressed this question using eastern white cedar (*Thuja occidentalis*) as a model species for declining to endangered long-lived tree species with mixed-mating system.

Results

We determined the SGS in two core and two peripheral populations of eastern white cedar from its Maritime Canadian eastern range using six nuclear microsatellite DNA markers. Significant SGS ranging from 15 m to 75 m distance classes was observed in the four studied populations. An analysis of combined four populations revealed significant positive SGS up to the 45 m distance class. The mean positive significant SGS observed in the peripheral populations was up to six times (up to 90 m) of that observed in the core populations (15 m). Spatial autocorrelation coefficients and correlograms of single and sub-sets of populations were statistically significant. The extent of within-population SGS was significantly negatively correlated with all genetic diversity parameters. Significant heterogeneity of within-population SGS was observed for 0-15 m and 61-90 m between core and peripheral populations. Average *Sp*, and gene flow distances were higher in peripheral (*Sp* = 0.023, $_g$ = 135 m) than in core (*Sp* = 0.014, $_g$ = 109 m) populations. However, the mean neighborhood size was higher in the core (*Nb* = 82) than in the peripheral (*Nb*= 48) populations.

Conclusion

Eastern white cedar populations have significant fine-scale genetic structure at short distances. Peripheral populations have several-folds higher within-population fine-scale genetic structure than core populations. Anthropogenic disturbances and population fragmentation presumably have significant effects on fine-scale genetic structure in eastern white cedar. Core populations have higher neighborhood size than peripheral populations, whereas gene flow distances are higher in peripheral than in core populations. The results of our study contribute to the knowledge of poorly-understood spatial genetic structure of core versus peripheral populations in plants. As well, the information is of significance for conservation of genetic resources of eastern white cedar and perhaps of other long-lived forest trees with mixed-mating system.

BACKGROUND

Recently, in addition to core, peripheral populations are also gaining importance for gene conservation, because under global climate change conditions they may possess genotypes of future adaptive potential. Plant populations tend to migrate further from their range periphery to ensure their survival by developing adaptability in a changing environmental condition due to global warming [1]. Genetic studies involving both core and peripheral populations are still limited[2]. Additionally, the available studies mainly focused on within-and among-population genetic diversity [2-4], and majority of them reported higher genetic diversity and lower population differentiation in core than in peripheral populations. Although still limited, there are increasing reports on SGS in fragmented versus continuous populations e.g., [5-7]. However, information on the comparison of SGS between core and peripheral populations of plant species is almost non-existent. The only published report on this aspect by Gapare & Aitken [8] reported significantly higher SGS in peripheral and random distribution of genotypes in core populations of predominantly outcrossing conifer, Sitka spruce (*Picea sitchensis*). The level of SGS within plant populations depends on seed and pollen dispersal, mating system, mode of reproduction, size of the founder population, species longevity, adult population density, effective population size, and micro environmental selection [9-13]. Anthropogenic disturbance and fragmentation of populations can also influence SGS by restricting gene flow and increasing local drift due to severe modification of forest environment and ecosystem, and density of populations [14,15]. Information on SGS levels in core and peripheral populations can unravel differential effect of the key evolutionary and ecological genetic processes operating in natural plant populations, such as gene dispersal, natural selection and genetic drift. Analysis of SGS is also important for understanding the population dynamics of a species as its response to micro-environmental variation and global climate change [16,17]. Furthermore, knowledge of SGS in natural populations is also important for developing scientifically-sound sampling strategies for genetic resource conservation and breeding purposes.

Higher SGS may be expected in peripheral than in core populations of long-lived plants, such as forest trees, owing to the differences in founding population, seed and pollen dispersal, adult plant density, inbreeding levels, effective population size (N_e), and disturbance and selection regimes. Restricted gene flow due to limited seed and pollen dispersal can promote mating among relatives which can result in isolation by distance within a population [18]. Therefore, a local genetic subdivision can develop SGS within populations. Selfing and other forms of inbreeding can limit gene dispersal, consequently influencing the spatial distribution of genetic variation within populations [16]. In contrast, generally, outcrossing decreases population subdivision by reducing mating among relatives and increasing N_e [10]. Therefore, higher level of SGS can be expected in inbreeding species as compared to outcrossing species. Vegetative clonal reproduction can significantly affect the SGS through the development of spatial clusters of the members of a clone[19]. Long-lived plants are expected to have low levels of SGS within their populations because of decreased population divergence due to increased N_e and decreased genetic drift over time [10]. SGS has been found to be inversely correlated with adult density within populations [20]. Additionally, the levels of SGS were found to be positively correlated with the level of fragmentation of plant populations e.g., [6,14,21].

Conifers are generally long-lived plants that are ecologically and economically important globally. They constitute a major component of the northern boreal and temperate forests, especially dominating the boreal forest which comprises one of the largest biomes in the worldhttp://www.borealforest.org/index.php?category=world_boreal_

forest&page=overview *webcite*. Most of the conifer tree species are characterized by predominantly (~90%) outcrossing mating system. Studies in conifers reported either no or weak levels of SGS within populations [21-32]. This is consistent with the conifers' characteristics of predominant outcrossing, wind-pollination and seed dispersal by wind. Except for two studies [24,32], all of the SGS studies in conifers are reported for *Pinus* or *Picea* species. Furthermore, with the exception of two [8,28], all other studies on SGS in conifers are based on only one or two populations. All of the SGS studies in conifers have been reported for predominantly outcrossing (≥90%) species, and there is no SGS information on plant species that have mixed-mating system with high selfing levels.

Eastern (or northern) white cedar (*Thuja occidentalis* L.) provides an ideal model species to study SGS in core versus peripheral populations of declining and long-lived plant species with mixed mating system. Eastern white cedar is a wind-pollinated, monoecious, evergreen tree species belonging to the Cypress family (Cupressaceae). It reproduces both sexually and vegetatively. Vegetative reproduction occurs mainly in swampy sites by layering [33]. Eastern white cedar is an ecologically and economically important tree species, highly suitable for many wood products. Since its foliage is a favorite winter food for deer, it is very important for wildlife conservation. It can reach an age of 400 years [34].

The natural distribution range of eastern white cedar extends from Manitoba east to Nova Scotia, and south into the Great Lakes Regions of the United States, east to New England, with southern disjunct populations in the Appalachians in North Carolina and Tennessee ([34]; Figure 1). Eastern white cedar reaches its eastern periphery in Nova Scotia and the Prince Edward Island (PEI), where it occurs in small, scattered and fragmented populations [34,35]. Historical records indicate that the range and number of populations of eastern white cedar have diminished due to logging, and clearing for agriculture over the past 400 years in Nova Scotia and PEI [33,35]. Presently, there are only 32 identified populations of eastern white cedar in Nova Scotia [34], which are highly fragmented. Due to its rarity in Nova Scotia, eastern white cedar has been designated as a Red (endangered) species in this province. In Prince Edward Island, occurrence of eastern white cedar is limited to only a few populations, mainly found in the western part of the Island (West of Summerside). Eastern white cedar is common in New Brunswick, where it forms its best development [35]. Thus, New Brunswick represents the core or central range of this species. However, here also eastern white cedar has been listed as a species of concern.

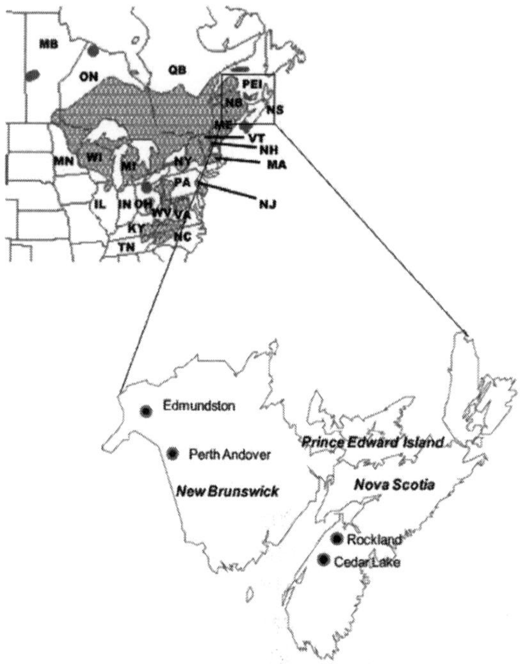

Figure 1: A map showing the geographical distribution [from:http://www.efloras.org/object_page.aspx?object_id=5261&flora_id=1] (above) of *Thuja occidentalis* and approximate locations of its four studied populations (below) (see also Table 1).

Like other cedar species (*Thuja plicata, T. orientalis*, [36-38]), eastern white cedar also possesses relatively low genetic diversity [39-41]. It is characterized by a mixed-mating system with high levels (mean 54%) of selfing rates [41,42]. Although a few genetic diversity studies have been conducted in the genus *Thuja* [36-41,43], to our best knowledge, there is no information available on SGS in any *Thuja* species. The only SGS study in the Cupressaceae family was done for one population of an outcrossing species, *Cryptomeria japonica* [32], which reported significant SGS at short distance classes (up to 60 m) based on nuclear microsatellites, but non-significant SGS for chloroplast microsatellites.

The aim of the present study was to determine within-population fine-scale genetic structure of eastern white cedar growing in core and peripheral populations. Our hypotheses were: (i) since eastern white cedar is a mixed-mating long-lived tree with high levels of selfing rate, and has capability to reproduce vegetatively, this species is expected to have some levels of SGS within population; (ii) the peripheral populations of eastern white cedar are expected to have higher levels of SGS than the core populations due to their lower population density; potentially restricted gene dispersal, smaller N_e, and higher anthropogenic fragmentation as compared to the core populations [9-13].

METHODS

Populations and Sampling

Four natural populations of eastern white cedar were sampled from two Maritime Provinces of Canada. Two of these were core (Edmundston and Perth Andover) populations from New Brunswick and other two were peripheral (Cedar Lake and Rockland) populations from Nova Scotia (Figure 1; Table 1). We restricted sampling of the core and peripheral populations to the same Maritime region of eastern Canada to avoid other regional and ecological effects confounding the results. The two populations sampled in New Brunswick are located in the central part of the species' range. Despite some harvesting activities, eastern white cedar populations remain intact with contiguous distribution in this province with no barrier to gene flow between populations. On the other hand, the two eastern white cedar populations sampled from Nova Scotia are disjunct, and highly fragmented and disturbed as they have experienced anthropogenic disturbances over a long period of time. It was not possible to find undisturbed pristine natural populations of eastern white cedar in both core and peripheral regions in eastern Canada. Also, fragmented core and contiguous peripheral populations could not be found in our study region. All of the sampled populations are over 100 years old. The populations sampled in Nova Scotia were from almost pure eastern white cedar stands, whereas the populations sampled from New Brunswick are mixed with other tree species, such as black spruce (*Picea mariana*) and balsam fir (*Abies balsamea*).

Table 1: Geographical location, sample size and stand density of four sampled populations of eastern white cedar

Province	Population	Code	**Range location**	**Latitude (N)**	**Longitude (W)**	**No. of individuals sampled**	**Stand density (Trees/ha)**
New Brunswick	Edmundston	ED	Core	47° 33′ 16″	68° 13′ 53″	100	46.72
New Brunswick	Perth Andover	PA	Core	46° 44′ 53″	67° 42′ 34″	100	58.82
Nova Scotia	Cedar Lake	CL	Peripheral	44° 50′ 22″	65° 01′ 48″	100	18.20
Nova Scotia	Rockland	RL	Peripheral	45° 00′ 35″	64° 42′ 10″	94	24.00

Pandey and Rajora BMC Evolutionary Biology 2012 12:48 doi: 10.1186/1471-2148-12-48

A plot consisting of ~ 100 contiguous eastern white cedar trees was established in each of the four sampled populations irrespective of the plot area (Figure 2). All of the eastern white cedar individuals were sampled within these plots. We chose this sampling strategy to keep the sample size equal among all four sampled populations, in order to minimize the effects of sample size confounding the comparison of SGS results between the populations. Spatial locations of each of the sampled trees were recorded as their geo-coordinates using a Geographical Positioning System (Garmin GPS MAP 60CSx). The spatial location maps of the sampled trees in each of the four populations are provided in Figure 2.

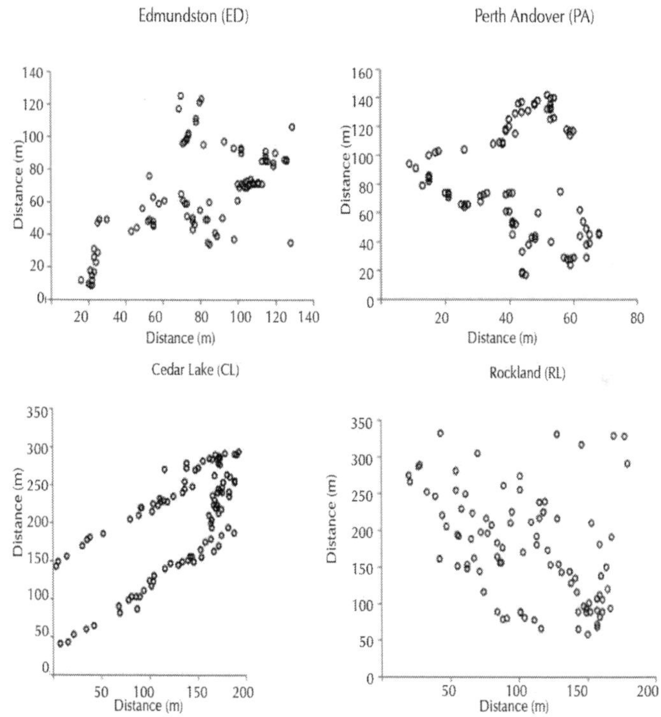

Figure 2: Spatial distribution of individual trees in each of the four studied eastern white cedar populations.

Small amounts of foliage samples were collected from each of the 394 sampled trees in plastic Ziplock bags containing silica gel. Upon arrival to the lab, the foliage samples were stored at -20°C until processed for DNA extraction.

DNA Extraction and Genotyping

Total genomic DNA was extracted from the needle samples of individual trees using the DNeasy Plant Mini Kit (Qiagen, USA). Microsatellite DNA markers of the nuclear genome were used for genotyping the sampled trees. A total of 40 microsatellite markers developed for the Cupressaceae species were tested in eastern white cedar. Of these, 37 were developed for *Thuja plicata* [44,45], [L.M. O'Connell & C.E. Ritland, Unpublished data; C.J. Newton, Unpublished data] and three for *Chameacyparis obtusa* [46]. Only six (TP1, TP2, TP4, TP6, TP11 and 91 G) of the 40 microsatellites yielded simple and scorable PCR (Polymerase Chain Reaction) products. These six microsatellites were optimized and used for genotyping individual eastern white cedar trees. PCR amplifications were carried out using 10 µl reaction mixtures containing 20 to 30 ng of genomic DNA, 1× PCR buffer, 1.5 mM $MgCl_2$, 0.20 mM of each dNTP, 0.50 pmol of each primer and 0.25 U *Taq* DNA polymerase (Invitrogen, USA). The epGradients Master Cycles (Eppendorf, Germany) were used for PCR amplification of the sample DNA, with the following thermocycling profile: initial denaturation step at 95°C for 5 min, followed by 33 cycles each of 95°C for 45 sec, annealing temperature at 55 to 65°C for 45 sec, 72°C for 45 sec and final extension step of 72°C for 5 min. PCR products were separated on LI-COR4300 genetic analyzers (LI-COR Inc. Lincoln, Nebraska, USA). The genotypes of individual trees were first determined by scoring the LI-COR gels with the SAGA GT program (LI-COR Inc. Lincoln, Nebraska, USA) and later verified by manual scoring.

Statistical Data Analysis

Genetic Diversity and N_e

The standard population genetic diversity parameters (number of alleles, observed and expected heterozygosity) and fixation index (*F*) were calculated using the program GENALEX [47]. Allelic richness (A_R) was estimated using the program FSTAT [48]. Genotypic diversity was determined by calculating observed and expected genotype additivity (GA_O and GA_E) as the sum of genotypes observed and expected under

Hardy-Weinberg equilibrium over the loci [49]. Deviations of genotypic frequencies from the Hardy-Weinberg expectation were tested using the Markov chain method (parameter values: dememorisation number = 10,000, number of batches = 300, number of iterations per batch = 5,000) in the GENEPOP program [50].

The effective population size (N_e) for each of the four populations was estimated using the maximum likelihood method of coalescent-theory and MIGRATE program [51]. At first, the values were estimated for each population using MIGRATE program and assuming an average microsatellite mutation rate of 10^{-3} per generation. The N_e was calculated as $/(4\times10.^{-3})$.

Spatial Distribution

Pattern of spatial distribution of individual trees in the four eastern white cedar populations were estimated using univariate second-order Ripley's [52]*K*-function for 10, 15, 20 and 30 m distance classes (t). The weighted edge correction [53] method was applied to correct the edge effect. Monte Carlo procedure was used to test the significance of deviation from a random distribution at 95% upper and lower confidence envelops, generating 10,000 replications. These calculations were performed using the SPPACK program [54].

Spatial Genetic Structure (SGS)

We used Smouse and Peakall's [12] spatial autocorrelation method to estimate the SGS in the four eastern white cedar populations employing the GENALEX 6.2 [47] program. This method determines spatial distribution of genetic variation based on a frequency-weighted average over all alleles and loci and correlation coefficients of genotypes in relation to their spatial distance. It allows for the covariance between alleles (see [12] for detailed description). In order to facilitate the comparison of SGS among the four populations, we used 15 m distance class as representative because three, out of the four populations showed maximum values of Ripley's *K* at this distance class. Similar approach was used by Oddou-Muratorio et al. [55] to select the distance class for SGS analysis in a tree species, *Sorbus torminalis*.

In addition to the estimation of multilocus SGS parameters for each of the four populations separately, SGS analyses were also performed for combined subset each of core (ED, and PA), and peripheral (CL and RL) populations, as well by combining all four populations to infer degrees of heterogeneity of SGS within and between subsets based on the method of Smouse et al. [56] using GENALEX 6.2 program. These analyses permit us to formally test whether differences in SGS between the subsets of populations are statistically significant. Since the analysis of SGS using combined multiple populations increases sample sizes, it increases statistical power of the analysis to detect the presence of SGS within populations. Heterogeneity tests for single and multi-populations correlograms were performed as described by Smouse et al. [56]. This method first performs the heterogeneity test of autocorrelation coefficient among individuals lag-by-lag using the pooled within-population correlograms. Then the single-lag tests are used to perform an analogous heterogeneity test of multilag correlograms (see Smouse et al. [56] for detailed description). Single class (t^2) and multi-class ( ) test criteria were used to test the heterogeneity of spatial autocorrelation (r) values between single and subsets of populations. Heterogeneity in SGS for single distance classes between any two populations or a subset of populations was determined by performing the squared paired- sample t test (t^2) [56]. SGS heterogeneity over all distance classes (a) in a population and (b) among populations was determined by using the test criteria [56] under the null hypotheses that (a) there is no significant spatial autocorrelation at any distance class, and that (b) the spatial correlograms from all populations are homogeneous. Statistical significance of all the analyses was performed by permutation test with 9,999 replications at the 95% upper and lower confidence intervals.

The correlations of the SGS parameters (significant spatial autocorrelation coefficient, r) with N_e(actual or log transformed), genetic diversity parameters and density of eastern white cedar trees within the sampled populations were determined by calculating the Pearson correlation coefficients (R), using the MINITAB 15 package.

Sp Statistics and Dispersal Estimates

In order to quantify SGS and to compute neighborhood sizes and gene dispersal distances in the four eastern white cedar populations, *Sp* statistic was estimated as follows: $Sp = -\hat{b}_F/(1-\hat{F}_1)$, where $-\hat{b}_F$ is the regression slope of the autocorrelogram and $\hat{F}_1$ is the mean kinship coefficient (*Fij*) [57] between individuals belonging to the first (0-15 m) distance class [21]. Using the *Sp*statistics, Wright's neighborhood size *Nb* ($Nb = 1/Sp$) was also estimated. *Nb* is defined as $Nb = 4\varpi D_{e\,g}^{\;2}$, where ${}_g^2$ is the axial variance of gene dispersal and D_e the effective population density. Estimates of gene dispersal distance (g) were determined assuming ratios of effective to census density, *De/D*, of 0.1 and 0.5 [7,21]. Estimates of $-\hat{b}_F$ and $\hat{F}_1$ were carried out using SPAGEDI program [58]. Statistical significance ($P \leq 0.05$) of the analyses was performed by permutation tests with 10,000 replications.

RESULTS

Genetic Diversity and Effective Population Size

In total, 95 alleles were observed at six microsatellite loci in four eastern white cedar populations. The core populations showed higher allelic as well as genotypic genetic diversity than the peripheral populations (Table 2). Also, the core populations possessed higher number of private alleles than the peripheral ones (Table 2). The two core populations from New Brunswick (ED and PA) showed very similar levels of genetic diversity. All four populations showed significant deviation from the Hardy-Weinberg equilibrium. The N_e was 432 and 910, with an average of 671 in the core populations, and 110 and 310, with an average of 210 in the peripheral populations (Table 2). As the objective of this study was to determine SGS, we do not focus on discussing genetic diversity results here. Results on genetic diversity and population structure of 13 eastern white cedar populations from the same region are presented and discussed elsewhere [59].

Table 2: Genetic diversity measures, fixation index and their (SE), and effective population size for four sampled populations of eastern white cedar

Pop type	Province	Population	A_T	A	A_E	A_R	A_P	H_O	H_E	GA_O	GA_E	F	N_e
Core	New Brunswick	ED	63.00	10.50 (1.86)	4.19 (1.15)	10.08 (2.09)	11	0.640 (0.057)	0.662 (0.079)	127 (3.84)	404 (7.43)	0.007 (0.062)	432
		PA	67.00	11.17 (2.63)	4.68 (1.51)	10.68 (2.51)	9	0.640 (0.087)	0.640 (0.103)	145 (4.53)	500 (9.03)	-0.025 (0.048)	910
Peripheral	Nova Scotia	CL	53.00	8.83 (2.02)	2.79 (0.44)	8.42 (2.15)	6	0.571 (0.070)	0.597 (0.060)	100 (3.12)	312 (6.69)	0.050 (0.034)	310
		RL	47.00	7.83 (2.17)	2.60 (0.48)	7.66 (2.27)	5	0.505 (0.111)	0.519 (0.110)	77 (2.94)	267 (7.07)	0.044 (0.048)	110
	Mean	Core	65.00	10.83 (2.24)	4.43 (1.33)	10.38 (2.30)	10	0.640 (0.072)	0.651 (0.091)	136.00 (4.19)	452 (8.23)	-0.009 (0.055)	671
		Peripheral	50.00	8.33 (2.10)	2.70 (0.46)	8.04 (2.21)	5.50	0.540 (0.091)	0.560 (0.085)	88.50 (3.03)	289.50 (6.88)	0.047 (0.041)	210
	Overall mean		57.50	9.58 (2.17)	3.57 (0.89)	9.21 (2.26)	7.75	0.590 (0.082)	0.600 (0.088)	112.25 (3.61)	370.75 (7.55)	0.019 (0.048)	440

A_T, total number of alleles; A, mean number of alleles per locus; A**E**, effective number of alleles; A_R, Allelic richness; A_P, number of private alleles; GA_O, observed genotype additivity (richness); GA_E, expected genotype additivity [49]; H**O**, observed heterozygosity; H**E**, expected heterozygosity; *F*, fixation index; N_e, effective population size. Values in parentheses are standard errors.

ED, Emundston; PA, Perth Andover; CL, Cedar Lake; RL, Rockland

Spatial Distribution

Results of spatial distribution pattern analysis using Ripley's *K*-function and the number of pairs of observation for 15 m distance class are presented in Figure 3, and Table 3, respectively. All four populations showed significant aggregation of individuals at the distance classes 10, 15, 20 and 30 m (results are shown only for 15 m distance class).

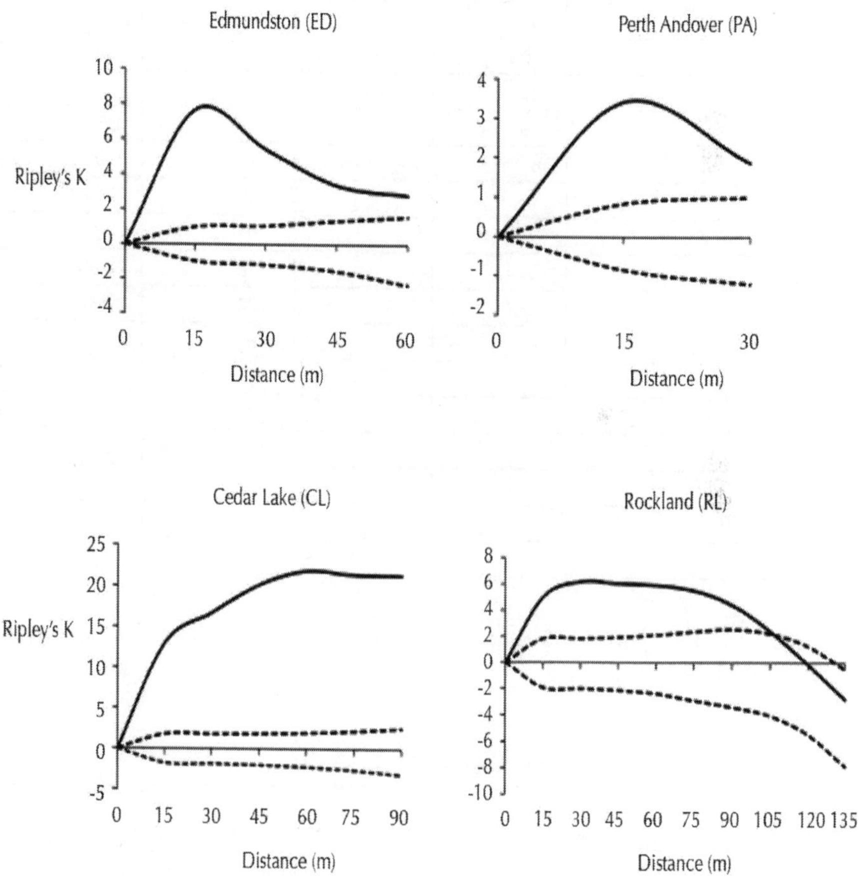

Figure 3: Patterns of spatial distribution of individual trees in four eastern white cedar populations using Ripley's *K* function (solid line), dashed lines represent 95% confidence envelope.

Table 3: Number of pairs of observation per distance class for the four eastern white cedar populations

Distance class (m)	Population			
	ED	**PA**	**CL**	**RL**
0-15	581	651	231	122
16-30	768	1014	345	262
31-45	919	700	430	327
46-60	1025	737	473	385
61-75	701	816	469	427
76-90	307	550	567	432
91-105	303	344	435	399
106-120	249	126	324	371
121-135	89	12	308	381
136-150	8		271	301
151-165			209	217
166-180			203	181
181-195			158	154
196-210			124	100
211-225			96	116
226-240			63	85
241-255			49	63
256-270			41	38
270-285			31	7
286-300			17	3
301-315			7	

ED, Emundston; PA, Perth Andover; CL, Cedar Lake; RL, Rockland

Pandey and Rajora BMC Evolutionary Biology 2012 12:48 doi: 10.1186/1471-2148-12-48

Spatial Genetic Structure (SGS)

Results of the multilocus spatial genetic structure (SGS) analysis based on Smouse & Peakall's[12] spatial autocorrelation coefficient (*r*) are presented in Figure 4. All four studied populations showed positive significant SGS in short distance classes ranging from 15 m in ED and PA to 75 m in RL (Figure 4). Analysis of the combined core (ED and PA) populations showed significant SGS up to 15 m, whereas analysis of the combined peripheral (CL and RL) populations showed significant SGS up to 90 m (Table 4). An overall significance test (ω) of spatial autocorrelations and correlograms revealed that all single or combined-populations were statistically significant (Table4).

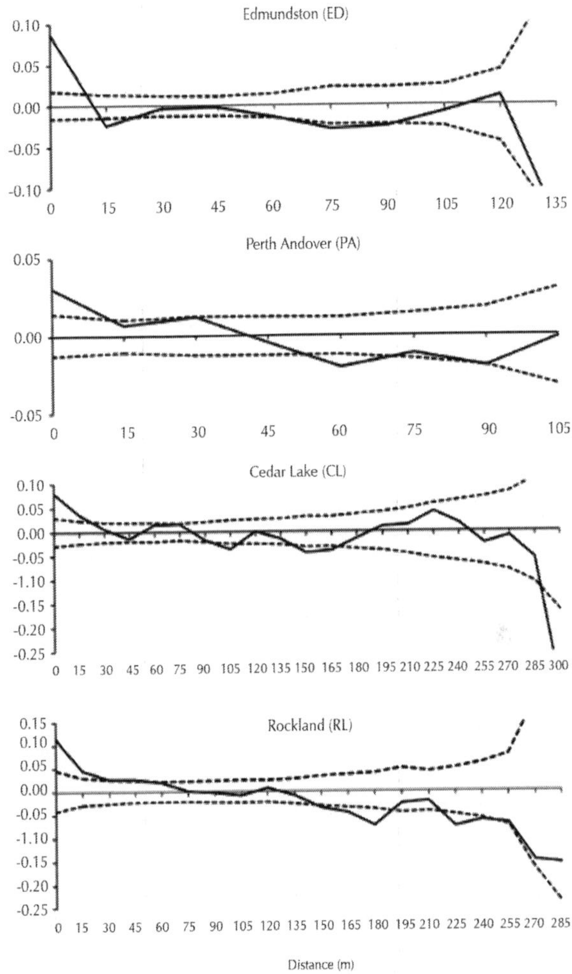

Figure 4: Correlograms of spatial autocorrelation coefficients (*r*) (solid lines) plotted against geographical distance classes, estimated for four eastern white cedar populations. Dashed lines represent upper and lower 95% confidence limits.

Table 4: Spatial autocorrelation coefficients (*r*), single-class squared pair-sample *t*-test (t^2) and multi-class and multi-population (ω) heterogeneity tests of spatial autocorrelation coefficients (*r*) for individual populations, population pairs and subsets of populations

Populations	***r* and t^2 values and their statistical significance in different distance classes**								**Overall multi-distance class or multi population correlogram significance (ω-test criterion)**
	0-15	16-30	31-45	46-60	61-75	76-90	91-105	106-120	
	Spatial autocorrelation coefficients (*r*)								
ED	0.087*	-0.024	-0.003	-0.002	-0.014	-0.029	-0.025	-0.008	75.35*
PA	0.030*	0.007	0.012	-0.005	-0.020	-0.012	-0.020	-0.002	70.49*
CL	0.081*	0.036*	0.006	-0.014	0.015	0.016	-0.017	-0.037	134.39*
RL	0.114*	0.045*	0.026*	0.026*	0.019*	0.001	-0.003	-0.009	147.32*
Core	0.058*	-0.007	0.003	-0.003	-0.017	-0.018	-0.022	-0.006	81.11*
Peripheral	0.093*	0.040*	0.016*	0.005*	0.017*	0.009*	-0.010	-0.022	154.80*
Four pops combined	0.066*	0.005	0.007*	0.000	-0.004	-0.003	-0.015	-0.016	65.65*
	Single-class squared pair-sample *t*-test (t^2)								
ED vs. PA	17.53*	8.44*	1.89	0.06	0.37	1.13	0.09	3.89	47.99*
ED vs. CL	0.10	17.13*	0.52	0.96	5.50*	8.03*	0.21	0.16	51.66*
ED vs. RL	1.34	18.83*	4.25*	4.70*	6.67*	3.07	1.71	0.02	57.02*
PA vs. CL	7.82*	4.54*	0.18	0.52	8.69*	4.31*	0.02	3.57	49.70*
PA vs. RL	12.76*	6.15*	0.93	5.05*	10.11*	0.72	1.03	4.00	62.34*
CL vs. RL	1.55	0.24	1.48	7.07*	0.07	1.20	0.85	0.16	28.08
Core vs. peripheral	4.72*	18.64	1.58	0.73	15.08*	6.83*	1.04	1.23	65.26*

ED, Edmundston; PA, Perth Andover; CL, Cedar Lake; RL, Rockland. * indicates significant values ($P \leq 0.05$)

Pandey and Rajora BMC Evolutionary Biology 2012 12:48, doi:10.1186/1471-2148-12-48

The maximum values of spatial autocorrelation coefficient (*r*) were observed for the first (0-15 m) distance class for individual or combined populations (Table 4). The *r* values for the first distance class showed great variation among the four populations ranging from 0.030 (PA population) to 0.114 (RL population) (Figure 4). Thus, *r in* RL was about four times of that in PA. The highest *r* for the combined peripheral populations (r = 0.093) was considerably higher than in the intact populations (r = 0.058).

Except for the population pair CL and RL, all pairs revealed significant heterogeneity in SGS between the populations based on the multi-class test criterion (ω) (Table 4). Significant heterogeneity in SGS was also observed between the core and peripheral populations over the 0-15 m and 61-90 m distances (up to 90 m) (Table 4).

Sp Statistics and Dispersal Estimates

Results of *Sp* statistics and dispersal estimates in the four eastern white cedar populations are presented in Table 5. *Sp* values were 0.009 (PA) and 0.018 (ED) in the core populations, and 0.016 (CL) and 0.031 (RL) in the peripheral populations. Average *Sp* value was higher in the peripheral (0.023) than in the core (0.014) populations Average value for Wright's neighborhood size (*Nb*) was higher in the core populations (81.57) as compared to the peripheral populations (48.04). Gene flow distances (σ_g) were 44 and 120 m in the core populations and 46 and 167 m in the peripheral populations (Table 5).

Table 5: Spatial genetic structure, Sp statistics and estimates of gene flow distances in four eastern white cedar populations

Region	Population	^F1	^F2	^F3	^F4	^F5	b-log (^bF)	Sp	Nb	$_g$(m) (De/D = 0.1)	$_g$(m) (De/D = 0.5)
Core	ED	0.047*	-0.013*	-0.001	-0.001	-0.008	-0.017*	0.018	56.73	98.29	43.96
	PA	0.015*	0.004	0.006	0.008	-0.013	-0.009*	0.009	106.41	119.98	53.66
Peripheral	CL	0.045*	0.023*	0.003	-0.008	-0.005	-0.015*	0.016	64.00	167.32	74.83
	RL	0.070*	0.027*	0.026*	0.015*	0.014*	-0.029*	0.031	32.08	103.14	46.13
Average core		0.031	0.004	0.003	0.004	-0.011	-0.013	0.014	81.57	109.14	48.81
Average peripheral		0.057	0.025	0.015	0.004	0.005	-0.022	0.023	48.04	135.23	60.48

^F1, ^F2, ^F3, ^F4 and ^F5: mean kinship coefficient (Fij) [57] between individuals belonging to the first (0-15 m), second (15-30 m), third (30- 45 m), fourth (45-60 m) and fifth (60-75 m) distance classes, respectively; b-log (^bF): Slope of the regression of kinship with ln(dist); Sp: statistic defined by the ratio _^bF/(1 - ^F1), where ^bF is the regression slope of the autocorrelogram; ^F1 is the mean kinship coefficient (Fij) between individuals belonging to the first (0-15 m) distance class [20]; Nb Wright's neighbourhood size; g, estimate of gene flow distance from Sp, assuming different ratios of effective (De) to census density (D); *, significant (P < 0.05)

ED, Edmundston; PA, Perth Andover; CL, Cedar Lake; RL, Rockland

Pandey and Rajora BMC Evolutionary Biology 2012 12:48 doi:10.1186/1471-2148-12-48

Correlations of SGS Parameters with Genetic Diversity, Stand Density, and Population *Ne*

There was negative correlation of the significant SGS (spatial autocorrelation coefficient *r*) with the population genetic diversity measures. The correlations (Pearson correlation coefficient R) of *r* with genetic diversity measures were as follows: mean number of alleles per locus (A) = -0.85 $P = 0.034$), observed heterozygosity (H_O) = - 0.89 ($P = 0.022$), expected heterozygosity (H_E) = -0.82 ($P = 0.044$), observed genotypic richness (GA_O) = -0.88 ($P = 0.024$) and expected genotypic richness (GA_E) = 08 $P = 0.040$). The correlation of eastern white cedar stand density with *r* a -0.69$P = 0.096$). The significant SGS *r* values showed negative high correlation (Pearson R = -0.65)wt*Ne*, although the correlation was not statistically significant ($P = 0.139$).

DISCUSSION

Spatial Genetic Structure in Eastern White Cedar

Our study demonstrates that eastern white cedar has significant within-population fine-scale genetic structure at short distances. This was clearly evident from significant positive SGS at ~15 m to ~75 m in four studied eastern white cedar populations; suggesting that individuals with similar alleles or genotypes were more often growing in proximity than expected by chance alone. The extent of the observed significant SGS depended upon the location of the populations (core versus peripheral). These results are discussed later in this section.

The extent of SGS for combined four eastern white cedar populations (up to 45 m) is more than two times of that observed in one 20-years old population of another mixed-mating (mean selfing rate of 27% [60]) species *Larix larcinia* (up to 20 m) [33]. The existence of SGS in *Larix larcinia*was explained primarily due to the effect of founding population. In contrast, predominantly-outcrossing species showed either weak or no within-population SGS e.g. [17,22,28]. Thus, mixed-mating conifer species tolerating high levels of selfing apparently have higher SGS within their populations than the predominantly outcrossing conifers.

Besides higher levels of selfing or inbreeding, limited gene flow and founding gene pool may be responsible for the observed short-distance SGS in eastern white cedar. Restricted gene flow due to limited pollen and seed dispersal contributes to local grouping of individuals with similar alleles or genotypes [10]. In eastern white cedar, both pollen and seeds are disseminated by wind. We are not aware of any information on the average distance of pollen dispersal in this species. However, in conifers it is well known that pollen is dispersed in enormous amounts over considerable distances [61,62]. Therefore, the restriction in gene flow due to limited pollen flow is less likely to happen in eastern white cedar, especially in the contiguous populations. However, in the peripheral populations, reduced gene flow may have contributed to the observed levels of fine-scale genetic structure as a result of fragmentation. Fowells [63] reported a seeding range of up to 60 m in eastern white cedar under normal conditions. Interestingly, the mean level of SGS observed in eastern white cedar (up to 45 m) falls within this range of the seeding distance. Therefore, restricted gene flow due to limited seed dispersal may be one of the main reasons for the observed short-distance SGS in the species.

Vegetative propagation through natural layering has been reported in eastern white cedar [33,64]. Since the individuals established from vegetative propagation have genetic constitution identical to the source tree and grow in proximity forming clumps, this may generate a significant SGS at short distance intervals. However, we did not find any concrete evidence for vegetative propagation contributing to the observed levels of SGS in any of the four study populations. We detected only four to eight individuals (4 to 8%) per population growing in proximity of each other with the same multilocus genotypes (data not shown). This level of multilocus genotype sharing may be due to vegetative reproduction, selfing, a combination of these factors or just by chance. However, our data does not support vegetative propagation contributing to the observed levels of SGS in eastern white cedar. Although, it is generally expected that vegetative reproduction is one of the major factors that may contribute to the occurrence of significant SGS in vegetatively reproducing species, Chung & Epperson [65] found little effect of vegetative reproduction on SGS in *Eurya emarginata*. They explained that the restriction in gene flow due to limited dispersal of seed and pollen could be the main reason for the observed SGS in the species.

Spatial Distribution Pattern of Trees

The analysis of spatial distribution pattern using Ripley's *K* function indicated significant aggregation of trees in all four populations of eastern white cedar. Within population spatial patterns of individual plants can be influenced by several factors such as variation in microsites, level of disturbance, seed dispersal patterns [56,66]. Parker et al. [66] and Trabaud et al. [67] reported a significant aggregation of tress in *Pinus clausa* and *Pinus halepensis* populations, respectively, and they explained that the observed aggregation could be due to either limited seed dispersal or recruitment of aggregated seedlings after stand disturbance due to fire. These factors may have contributed to the observed spatial pattern of individual trees in eastern white cedar populations. The effect of limited seed dispersal on the significant aggregation of individual trees may be evident from the significant SGS (15-75 m) observed in the four populations. Although we are not aware of any records of fire in the studied populations, past occurrence of fire on the studied populations cannot be ruled out because fire is one of the main agents of natural disturbance in boreal forest [68].

SGS in Core versus Peripheral Populations

The mean significant SGS in the peripheral populations (up to 90 m) was up to six times of that was observed in the core (15 m) populations. The difference in SGS between the core and peripheral populations was statistically significant ($\omega = 65.26$, $P = 0.001$). Both Nova Scotia populations (CL and RL) showed the highest within-population

SGS, and the lowest genetic diversity, stand density and N_e. These populations are highly fragmented and affected by human disturbance for over 400 years ever since the arrival of European settlers in the Province. The several-folds higher SGS levels observed in the two marginal and highly fragmented populations from Nova Scotia is likely due to reduced stand density, gene dispersal, genetic diversity and N_e observed in these populations (Tables 1 and 2) as well as increased founder effect and genetic drift due to anthropogenic fragmentation and other disturbances. Thus, the magnitude of evolutionary forces operating in core versus peripheral populations is probably different. It was not possible to dissect out the effects of anthropogenic fragmentation and disturbances on SGS in the peripheral populations from that of their peripheral location. Comparisons of SGS between pristine core and pristine peripheral populations would have been ideal. However, in the eastern Maritime Canadian range of eastern white cedar where our study was focused, no peripheral population could be found that is not anthropogenically fragmented and disturbed. Thus, fragmentation and disturbance may also have contributed to higher SGS observed in the peripheral populations. The negative correlations of SGS with stand density, genetic diversity parameters and N_e are consistent with the population genetic theories and expectations. The negative correlation between genetic diversity and SGS could be due to the combined effect of founder effects and reduced stand density.

The spatial distribution of genotypes was found to be random in the core populations of Sitka spruce, whereas significant positive SGS up to 500 m was observed in the peripheral populations of this species [8]. In eastern white cedar, we observed mean significant positive SGS up to 90 m (combined populations) in the peripheral and up to 15 m in the core populations. Thus, the magnitude of impact of ecological and evolutionary factors on SGS in peripheral populations of Sitka spruce appears to be much higher than in eastern white cedar. Sitka spruce is a predominantly outcrossing species. The extent of differences observed in within-population SGS in core versus peripheral populations between these species may be related to the differences in their mating system. The mixed-mating species with relatively low inherent genetic diversity, such as eastern white cedar, can perhaps tolerate better negative effects (increased inbreeding, decreased genetic diversity and N_e) than the predominantly outcrossing species with relatively high genetic diversity, such as Sitka spruce. However, this aspect remains to be further tested in more details globally.

SP Statistics and Gene Dispersal Distance

In agreement with the SGS results estimated using Smouse and Peakall's [12] autocorrelation coefficient method, average value for *Sp* statistic was also higher in peripheral than in core populations of eastern white cedar. Vekemans and Hardy [20] reported significantly higher levels of *Sp* value in populations with low density as compared to the populations with high density. Since the density of trees in the peripheral populations was much lower than in the core ones, the effect of density on the results of *Sp* cannot be ruled out. Except in the population PA, *Sp* values observed in the other three populations of eastern white cedar were higher than the average value observed in five wind-dispersed temperate trees (*Sp* = 0.012) and 18 outcrossing plant species[20]. However, it was lower in eastern white cedar as compared to the average value (*Sp* = 0.037) of seven mixed mating species [20]. De-Lucas et al. [7] reported higher *Sp* values in fragmented than in continuous populations of another conifer species, *Pinus pinaster*, and they explained that the lower effective population size due to few reproductive individuals in fragmented populations may have resulted in restricted gene flow via seed and increased the level of SGS in the populations. This explanation may also be applicable to eastern white cedar because the peripheral populations were highly affected by anthropogenic fragmentation. The average gene flow distance in eastern white cedar was higher in peripheral populations than in the core ones. Higher gene flow distances would be expected in fragmented and peripheral populations than in core and intact populations because of lower tree density in peripheral populations resulting in lower obstruction for pollen movement [20,69], which is the case in peripheral populations of eastern white cedar. In agreement with our results, Vekemans and Hardy [20] also reported higher gene flow distances in low density populations than in populations with high density.

Evolutionary Potential and Genetic Resource Conservation Implications

The results of genetic diversity, *N*e and SGS comparisons in core versus peripheral populations suggest that the evolutionary potential of the studied peripheral eastern white cedar populations may have been adversely affected. The results of our study have significance and implications for conservation and management of genetic resources of eastern white cedar - a declining to endangered species in Atlantic Canada. For *ex-situ* conservation via seed and vegetative propagule collection, a minimum distance of 75 meters in peripheral and 20 m in core populations is advisable to capture maximum genetic diversity and avoid sampling genetically-related trees. The core populations in New Brunswick should be conserved *in-situ*. Both of the peripheral Nova Scotia populations have suffered genetic degradation as a result of extensive human disturbance and edge effect. Here, both *ex-situ* and *in-situ* genetic resource conservation efforts supplemented with genetic restoration will be required given that eastern white cedar is a red-listed species in Nova Scotia. The results of our study are relevant for within-population sampling of individuals for population genetic studies in this species. The results may also be applicable for conservation and management of genetic resources of other long-lived, declining species with similar mixed-mating characteristics.

CONCLUSIONS

In conclusion, our results clearly demonstrate that the studied eastern white cedar populations have significant fine-scale genetic structure at short distances. Peripheral populations have several-folds higher within-population fine-scale genetic structure than core populations. Vegetative propagation does not apparently contribute to SGS in eastern white cedar. Anthropogenic disturbances and fragmentation presumably have significant effects on fine-scale genetic structure in eastern white cedar. The magnitude of evolutionary forces such as founder effect, genetic drift, gene flow and selection, operating in core versus peripheral populations is likely to be different. This may be the first report comparing SGS in intact core versus fragmented peripheral populations of a long-lived, mixed-mating declining plant. The results of our study contribute to the knowledge of poorly-understood spatial genetic structure of core versus peripheral populations in plants.

AUTHORS' CONTRIBUTIONS

Both authors have contributed significantly to the work reported in the manuscript. MP was a Postdoctoral Fellow, who generated 50% of the eastern white cedar genotype data, performed data analysis, and drafted the manuscript. OPR is the principal investigator of the project and contributed to the conception of the study and its experimental design, provided overall guidance and direction and funding, and revised the manuscript. Both authors have read and approved the final manuscript.

ACKNOWLEDGEMENTS

We thank Tony Duke of the Nova Scotia Department of Natural Resources for drawing our attention to eastern white cedar conservation genetic issues. We also thank Lawrence Benjamin of Nova Scotia Department of Natural Resources, Michael H. McDonald of New Brunswick Department of Natural Resources and Brenda Tse for their help with the field work and Brenda Tse and Sandra Tucker for their laboratory assistance. Our sincere thanks go to Dr. Lisa O'Connell, who conducted field sampling, DNA extraction and genotyping work for two of the four populations sampled in this study. The research was funded by the research grants provided by the Nova Scotia Department of Natural Resources, PEI Model Forest Network Partnerships Ltd., Canada Research Chair Program (CRC950-201869) funds and the Natural Sciences and Engineering Research Council of Canada Discovery Grant (RGPIN 170651) to O.P. Rajora.

REFERENCES

1. Aitken SN, Yeaman S, Holliday JA, Wang T, Curtis-McLane S: Adaptation, migration or extirpation: climate change outcomes for tree populations. *Evol Appl* 2008, 1:95-111.

2. Eckert CG, Samis KE, Lougheed SC: Genetic variation across species' geographical ranges: the central-marginal hypothesis and beyond. *Mol Ecol* 2008, 17:1170-118
3. Lesica P, Allendorf FW: When are peripheral populations valuable for conservation? *Conserv Biol* 1995, 9:753-760.
4. Hampe A, Petit RJ: Conserving biodiversity under climate change: the rear edge matters. *Ecol Letters* 2005, 8:461-467.
5. Williams DA, Wang Y, Borchetta M, Gaines MS: Genetic diversity and spatial structure of a keystone species in fragmented pine rockland habitat. *Biol Conserv* 2007, 138:256-268.
6. Born C, Hardy OJ, Chevalier MH, Ossari S, Attéke EJW, Hossaert-Mckey M: Small-scale spatial genetic structure in the central African rainforest tree species *Aucoumea klaineana*: a stepwise approach to infer the impact of limited gene dispersal, population history and habitat fragmentation. *Mol Ecol* 2008, 17:2041-2050.
7. De-Lucas AI, Gonza'lez-Marti'nez SC, Vendramin GG, Hidalgo E, Heuertz M: Spatial genetic structure in continuous and fragmented populations of *Pinus pinaster*Aiton. *Mol Ecol* 2009, 18:4564-4576.
8. Gapare WJ, Aitken SN: Strong spatial genetic structure in peripheral but not core populations of Sitka spruce [*Picea sitchensis* (Bong.) Carr.]. *Mol Ecol* 2005, 14:2659-2667.
9. Jain SK: Population structure and the effects of breeding system. In *Plant Genetic Resources: Today and Tomorrow*. Edited by Frankel O, Hawkes JG. Cambridge Univ. Press, Cambridge, UK; 1975:15-36.
10. Loveless MD, Hamrick JL: Ecological determinants of genetic structure in plant populations. *Ann Rev Ecol Syst* 1984, 15:65-95.
11. Hamrick JL, Nason JD: Consequence of dispersal in plants. In *Population Dynamics in Ecological Space and Time*. Edited by Rhodes OE, Ronald KC, Smith MH. The University of Chicago Press, Chicago, USA; 1996:203-235.
12. Smouse PE, Peakall R: Spatial autocorrelation analysis of individual multiallele and multilocus genetic structure. *Heredity* 1999, 82:561-573.

13. Kalisz S, Nason JD, Hanzawa FM, Tonsor SJ: Spatial population genetic structure in*Trillium grandiflorum*: the roles of dispersal, mating, history and selection. *Evol Biol* 2001, 55:1560-1568.
14. Young AG, Merriam HG: Effects of forest fragmentation on the spatial genetic structure of *Acer saccharum* Marsh (sugar maple) populations. *Heredity* 1994, 72:201-208.
15. Jump AS, Peñuelas J: Genetic effects of chronic habitat fragmentation in a wind-pollinated tree. *Proc Natl Acad Sci* 2006, 103:8096-8100.
16. Epperson BK: Plant dispersal, neighbourhood size and isolation by distance. *Mol Ecol* 2007, 16:3854-3865.
17. Troupin D, Nathan R, Vendramin GG: Analysis of spatial genetic structure in an expanding *Pinus halepensis* population reveals development of fine-scale genetic clustering over time. *Mol Ecol* 2006, 15:3617-3630.
18. Wright S: Isolation by distance. *Genetics* 1943, 28:114-138. PubMed Abstract | Publisher Full Text |PubMed Central Full Text
19. Reusch TBH, Hukriede W, Stam WT, Olsen JL: Differentiating between clonal growth and limited gene flow using spatial autocorrelation of microsatellites. *Heredity* 1999, 83:120-126.
20. Vekemans X, Hardy OJ: New insights from fine-scale spatial genetic structure analyses in plant populations. *Mol Ecol* 2004, 13:921-934.
21. Boyle TJB, Liengsiri C, Piewluang C: Genetic structure of black spruce on two contrasting sites. *Heredity* 1990, 65:393-399.
22. Epperson BK, Allard RW: Spatial autocorrelation analysis of the distribution of genotypes within populations of lodgepole pine. *Genetics* 1989, 121:369-377.
23. Knowles P: Spatial genetic structure within two natural stands of black spruce (*Picea mariana* (Mill.) B. S. P.). *Silvae Genet* 1991, 40:13-19.
24. Knowles P, Perry DJ, Foster HA: Spatial genetic structure in two tamarack (*Larix laricina* (Du Roi) K. Koch) populations with differing establishment histories. *Evolution* 1992, 46:572-576.
25. Xie CY, Knowles P: Spatial genetic substructure within natural populations of jack pine (*Pinus banksiana*). *Can J Bot* 1991, 69:547-551.

26. Leonardi S, Raddi S, Borghetti M: Spatial autocorrelation of allozyme traits in a Norway spruce (*Picea abies*) population. *Can J For Res* 1996, 26:63-71.
27. Epperson BK, Chung MG: Spatial genetic structure of allozyme polymorphisms within populations of *Pinus strobus* (Pinaceae). *Am J Bot* 2001, 88:1006-1010.
28. Parker KC, Hamrick JL, Parker AJ, Nason JD: Fine-scale genetic structure in *Pinus clausa* (Pinaceae) populations: effects of disturbance history. *Heredity* 2001, 87:99-113.
29. Bucci G, Menozzi P: Spatial autocorrelation and linkage of Mendelian RAPD markers in a population of *Picea abies* Karst. *Mol Ecol* 2002, 11:305-315.
30. González-Martínez SC, Gerber S, Cervera MT, Martínez-Zapater JM, Gil L, Alía R: Seed gene flow and fine-scale structure in a Mediterranean pine (*Pinus pinaster* Ait.) using nuclear microsatellite markers. *Theor Appl Genet* 2002, 104:1290-129.
31. Marquardt PE, Epperson BK: Spatial and population genetic structure of microsatellites in white pine. *Mol Ecol* 2004, 13:3305-3315.
32. Takahashi T, Tani N, Niiyama K, Yoshida S, Taira H, Tsumura Y: Genetic succession and spatial genetic structure in a natural old growth *Cryptomeria japonica* forest revealed by nuclear and chloroplast microsatellite markers. *For Ecol Manage* 2008, 255:2820-2828.
33. Newell RE: *Provincial (Nova Scotia) Status report on northern white cedar Thuja occidentalis prepared for the Nova Scotia Department of Natural Resources*. 2005.
34. Jhonston WF: *Thuja occidentalis* L.: northern white cedar. In *Silvics of North America. 1. Conifers*. Edited by Burns RM, Honkala BH. US Department of Agriculture Forest Service Handbook 654, Washington DC, USA; 1990:580-589.
35. Bentley PA, Smith EC: A study of northern white cedar and jack pine in Nova Scotia. *Proc NS Inst Sci* 1959, 24:376-398.
36. Xie CY, Dancik BP, Yeh FC: The mating system in natural populations of *Thuja orientalis*. *Can J For Res* 1991, 21:333-339.
37. Xie CY, Dancik BP, Yeh FC: Genetic structure of *Thuja orientalis*. *Biochem Syst Ecol* 1992, 20:433-441.

38. El-Kassaby YA, Russell J, Ritland K: Mixed mating in an experimental populations of western red cedar, *Thuja plicata. J Hered* 1994, 85:227-231.
39. Perry DJ, Knowles P, Yeh FC: Allozyme variation of *Thuja occidentalis* L. in Northwestern Ontario. *Biochem Syst Ecol* 1990, 18:111-115.
40. Matthes-Sears U, Mattes-Sears SC, Larson DW: Sources of allozymic variation in*Thuja occidentalis* in southern Ontario, Canada. *Silvae Genet* 1991, 40:100-105.
41. Lamy S, Bouchard A, Simon J-P: Genetic structure, variability and mating system in eastern white cedar (*Thuja occidentalis*) populations of recent origin in an agricultural landscape in southern Québec. *Can J For Res* 1999, 29:1383-1392.
42. Perry DJ, Knowles P: Evidence of high self-fertilization in natural populations of eastern white cedar (*Thuja occidentalis*). *Can J Bot* 1990, 68:663-668.
43. O'Connell LM, Ritland K, Thompson SL: Patterns of post-glacial colonization by western redcedar (*Thuja plicata*, Cupressaceae) as revealed by microsatellite markers. *Botany* 2008, 86:194-203.
44. O'Connell LM, Ritland CE: Characterization of microsatellite loci in western red cedar (*Thuja plicata*). *Mol Ecol* 2001, 9:1920-1922.
45. White E, Hunter J, Dubetz C, Brost R, Bratton A, Edes S, Sahota R: Microsatellite markers for individual tree genotyping: application in forest crime prosecutions. *J Chem Tech Biotech* 2000, 70:923-926.
46. Nakao Y, Iwata H, Matsumoto A, Tsumura Y, Tomaru N: Highly polymorphic microsatellite markers in *Chamaecyparis obtusa*. *Can J For Res* 2001, 31:2248-2251.
47. Peakall R, Smouse PE: GENALEX 6: genetic analysis in Excel. Population genetic software for teaching and research. *Mol Ecol Notes* 2006, 6:288-295.
48. Gaudet J: FSTAT version 1.2: a computer program to calculate F-statistics. *J Hered* 1995, 86:485-486.
49. Rajora OP, Rahman MH, Buchert GP, Danick BP: Microsatellite DNA analysis of genetic effects of harvesting in old-growth

eastern white pine (*Pinus storbus*) in Ontario. *Mol Ecol* 2000, 9:339-348. PubMed Abstract | Publisher Full Text

50. Raymond M, Rousset F: GENEPOP version 1.2: population genetics software for the exact tests and ecumenicism. *J Hered* 1995, 86:248-249.
51. Beerli P, Felsenstein J: Maximum likelihood estimation of migration rates and population numbers of two populations using a coalescent approach. *Genetics* 1999, 152:763-773.
52. Ripley BD: The second-order analysis of stationary point processes. *J Appl Prob* 1976, 13:255-266.
53. Ripley BD: Modelling spatial patterns. *J R Stat Soc* 1977, B39:172-212.
54. Perry GLW: SpPack: spatial point pattern analysis in Excel using Visual Basic for Applications (VBA). *Environ Model Soft* 2004, 19:559-569.
55. Oddou-Muratorio S, Demesure-Musch B, Pelissier R, Gouyon PH: Impacts of gene flow and logging history on the local genetic structure of a scattered tree species,*Sorbus torminalis* L. Crantz. *Mol Ecol* 2004, 13:3689-3702.
56. Smouse PE, Peakall R, Gonzales E: A heterogeneity test for fine-scale genetic structure. *Mol Ecol* 2008, 17:3389-3400.
57. Loiselle BA, Sork VL, Nason J, Graham C: Spatial genetic-structure of a tropical understory shrub, Psychotria officinalis (Rubiaceae). *Am J Bot* 1995, 82:1420-145
58. Hardy OJ, Vekemans X: SPAGEDi: a versatile computer program to analyse spatial genetic structure at the individual or population levels. *Mol Ecol Notes* 2002, 2:618-620.
59. Pandey M, Rajora OP: Genetic diversity and differentiation of core versus peripheral populations of eastern white cedar, *Thuja occidentalis L.* (Cupressaceae). *Am J Botany* 2012, 99:690-699.
60. Knowles P, Furnier GR, Aleksiuk MA, Perry DJ: Significant levels of self-fertilization in natural populations of tamarack. *Can J Bot* 1987, 65:1087-1091.
61. Levin DA, Kerster HW: Gene flow in seed plants. In *In Evolutionary Biology 7*. Edited by Dobzhansky T, Hecht MT, Steere WC. New York: Plenum Press; 1974:139-220.

62. Schuster WSF, Mitton JB: Paternity and gene dispersal in limber pine (*Pinus flexilis*James). *Heredity* 2000, 84:348-361.

63. Fowells HA: *Silvics of forest trees of the United States. U. S. Department of Agriculture*. Agriculture Handbook 271; 1965.

64. De Blois S, Bouchard A: Dynamics of *Thuja occidentalis* in an agricultural landscape of southern Quebec. *J Veg Sci* 1995, 6:531-542.

65. Chung MG, Epperson BK: Clonal and spatial genetic structure in *Eurya emarginata*(Theaceae). *Heredity* 2000, 84:170-177.

66. Parker KC, Parker AJ, Beaty M, Fuller MM, Faust TD: Population structure and spatial pattern of two coastal populations of Ocala sand pine (*Pinus clausa* (Chapm. ex Engelm.) Vasey ex Sarg. var. clausa D. B. Ward). *J Torr Bot Soc* 1997, 124:22-33.

67. Trabaud L, Michels C, Grosman J: Recovery of burnt *Pinus halepensis* Mill. forests. II. Pine reconstitution after wildfire. *For Ecol Manage* 1985, 113:67-179.

68. McCullough DG, Werner RA, Neumann D: Fire and insects in northern and boreal forest ecosystems of North America. *Ann Rev Entomol* 1998, 43:107-127.

69. O'Connell L, Mosseler A, Rajora OP: Extensive long-distance pollen dispersal in a fragmented landscape maintains genetic diversity in white spruce.

Chapter 5

Genetic Fragmentation in India's Third Longest River System, the Narmada

Gulab D Khedkar[1,3], Rahul Jamdade[1], Amol Kalyankar[1], Anita Tiknaik[1], Tetsuzan Benny Ron[2] and David Haymer[3]

[1]Paul Hebert Centre for DNA Barcoding and Biodiversity Studies, Dr. Babasaheb Ambedkar Marathwada University, Aurangabad 431004, India

[2]College of Tropical Agriculture and Human Resources, Department of Human Nutrition, Food and Animal Sciences, University of Hawaii, Honolulu, HI, USA

[3]Department of Cell and Molecular Biology, University of Hawaii, 1960 East-West Rd, Honolulu, HI 96822, USA

ABSTRACT

India's third longest river, the Narmada, is studied here for the potential effects on native fish populations of river fragmentation due to various barriers including dams and a waterfall. The species we studied include a cyprinid fish, *Catla catla,* and a mastacembelid, *Mastacembelus armatus*, both of which are found in the Narmada. Our goal was to use DNA sequence information from the D-loop region of the mitochondrial DNA to explore how this fragmentation could impact the genetic structure of these fish populations. Our results clearly show that these barriers can contribute to the fragmentation of the genetic structure of these fish communities, Furthermore, these barriers enhance the effects of natural isolation by distance and the asymmetry of dispersal flows. This may be a slow process, but it can create significant isolation and result in genetic disparity. In particular, populations furthest upstream having low migration rates could be even more subject to genetic impoverishment. This study serves as a first report of its kind for a river system on the Indian subcontinent. The results of this study also emphasize the need for appropriate attention towards the creation of fish passages across the dams and weirs that could help in maintaining biodiversity.

INTRODUCTION

India is endowed with diverse aquatic habitats due in part to a unique geological history, highly diverse physiography, a monsoon climate and high biotic diversity. The aquatic habitats include an extensive network of rivers and streams made up of medium and minor river systems (Rao 1975). The majority of these are perennial rivers with large seasonal variation in their flows. In India, over the past 60 years, the landscape of many rivers and streams has also been changed artificially. In response to the growing need of water for agriculture, industrialization and domestic use, many dams and reservoirs are constructed (Khedkar et al. 2014a). Water flow is now highly regulated and is often stored in reservoirs that impound nearly all medium and large rivers. Dams provide benefits in terms of flexibility to use water when it is needed for irrigation, generation of electricity, and other purposes, but have ecological costs as well.

The fragmentation of rivers due to dams or barriers may adversely affect fish populations by diminishing natural habitats required for all life stages and interfering with migration between populations. This in turn may lead to reduced population size, loss of genetic diversity, inbreeding and possible species extinctions. Dams may diminish natural habitats in rivers (Dynesius and Nilsson 1994) and can act as barriers that interfere with migration between populations, even in linked river segments. These barriers may also inhibit recolonization by neighbouring populations when local extinctions occur. River segments altered by dams can also lead to environmental disparities in terms of the temperature regimes, nutrient levels, substrate sizes, organic matter transport, the availability of lotic and lentic habitats, and the overall flow regimes (Ward and Stanford 1983). As a result, aquatic communities may experience altered seasonal movements, loss of genetic diversity, reduced population sizes and inbreeding (Ellstrand and Elam 1993; Lynch et al. 1995; Antunes Antunes et al. 2006). These alterations can lead directly to population extinctions (Fagan 2002).

Because the effects of such artificial alterations may have disproportionately greater impacts on the ability of migratory species to complete their entire life cycle, a number of studies have been directed toward recognizing the effects of river fragmentation due to dams on migratory fish populations (Dunham et al. 1997; Neraas and Spruell 2001; Morita and Yamamoto 2002). In addition, understanding the consequences of fragmentation may also be important for non-migratory species, such as cyprinids or mastacembelids, that have low clearing capacities to overcome the obstacle. Specifically it has been shown that although the entire fish community can be affected by weirs (Miranda et al. 2005; Poulet 2007), cyprinid species often make up a major portion of such river communities (Winfield and Townsend 1991).

Moreover, many cyprinid species are known to have to cover considerable distances for reproduction and feeding (McKeown 1984; Northcote 1984). One can hypothesise that such species would be sensitive to obstacles (Bainbridge 1958; Ovidio and Philippart 2002) and would be at great risk of population decline and loss of local genetic diversity (Knaepkens et al. 2004). However, relatively few studies have explored the impacts of river fragmentation due to dams and weirs on cyprinid populations. Of the studies done, many have been limited to small-scale observations such as the effects of single obstacles or have been assessed on individual-scale processes such as survival and/or migration only (Lucas and Frear 1997; Ovidio and Philippart2002).

In this study we used a broad scale approach to analyze the impact of fragmentation of a river due to a combination of factors including dams and natural barriers on populations of a cyprinid fish,*Catla catla*, and a mastacembelid fish, *Mastacembelus armatus*.

C. catla was selected as a model since it is a cyprinid, and although not considered to be migratory, is known to be able to cover distances of approximately 15-20 km against water current to spawn. *C. catla* is a ubiquitous species, and based on its size (commonly from 5–30 cm), surface dwelling habitat and its swimming ability, it can also be considered to be representative of many other cyprinids inhabiting Indian rivers (Jhingran 1968). Another species, *M. armatus,* was also selected for this study since it is an eel like species with a bottom dwelling habit (Rainboth1996; Vidthayanon 2002), is very common in Indian rivers and relatively easy to sample. *M. armatus* is not subject to manipulation as a source of material for hatchery based seed production and stocking, factors known to interfere with natural processes (Froese and Binohlan 2000).

We used DNA sequence information from the D-loop region of the mitochondrial DNA to identify markers to differentiate populations (Wilson and Cann 1985; Bremer et al. 1996; Nyakaana et al.2002; Sato et al. 2004; Khedkar et al. 2013) and to explore how both artificial and natural fragmentation of river habitats could impact the genetic structure of cyprinid and mastacembelid populations in the Narmada river in India. Our specific objectives were to evaluate the extent to which fragmentation could (i) prompt population differentiation at the genetic level, (ii) to identify the gross effects of multiple obstacles on the population's genetic structure in the upstream–downstream riverine regions; and (iii) to assess the effects of dams and a natural water fall on dispersion.

MATERIALS AND METHODS

Study Area

India's third longest river, known as the Narmada, is 1332 km long and covers a drainage basin totaling approximately 98,796 km^2. This study considers only the main stretch of the Narmada river and excludes its tributaries. Most of the study sites selected were dominated by cyprinid and mastacembelid fish. The Narmada is fragmented from upstream to downstream by several dams including the Bargi hydroelectric dam (69 m), the Indira Sagar hydroelectric dam (92 m) and Sardar Sarovar hydroelectric dam (136.5 m). In addition a series of six major natural waterfalls in the Bhedaghat area (~30 m height) occur along this part of the river (Table 1, Figure 1). Both the natural and artificial barriers impact the water flow and can potentially limit both upstream and downstream dispersion of the fish.

Table 1: Details of the sampling stations on the Narmada River

Sampling station	Geographic position			Distance from origin of the river (Km)	Distance from upstream sampling station (Km)	Distance from downstream sampling station (Km)	Presence of barrier up-stream and type	Presence of barrier down-stream and type
	Latitude	Longitude	Elevation (m)					
Dindori	22°56′32″N	81°04′34″E	2203	62	--	272	--	Bargi Hydroelectric dam (69 m high).
Jabalpur	23°04′55″N	79°54′12″E	1319	334	272	310	Bargi Hydroelectric dam (69 m high).	Series of 6 Natural Waterfalls (30 m high).
Hoshangabad	22°45′52″N	77°44′18″E	996	644	310	274	Series of 6 Natural Waterfalls (30 m high).	Indira Sagar Hydroelectric dam (92 m high).
Mortakka	22°09′03″N	75°29′54″E	545	918	274	280	Indira Sagar Hydroelectric dam (92 m high).	Sardar Sarovar Hydroelectric dam (136.5 m high).
Rajpipla	21°51′48″N	73°30′17″E	113	1198	280	105	Sardar Sarovar Hydroelectric dam (136.5 m high).	--
Bharuch	21°42′05″N	73°00′45″E	28	1303	105	--	--	--

Khedkar *et al.* *SpringerPlus* 2014 **3**:385 doi:10.1186/2193-1801-3-385

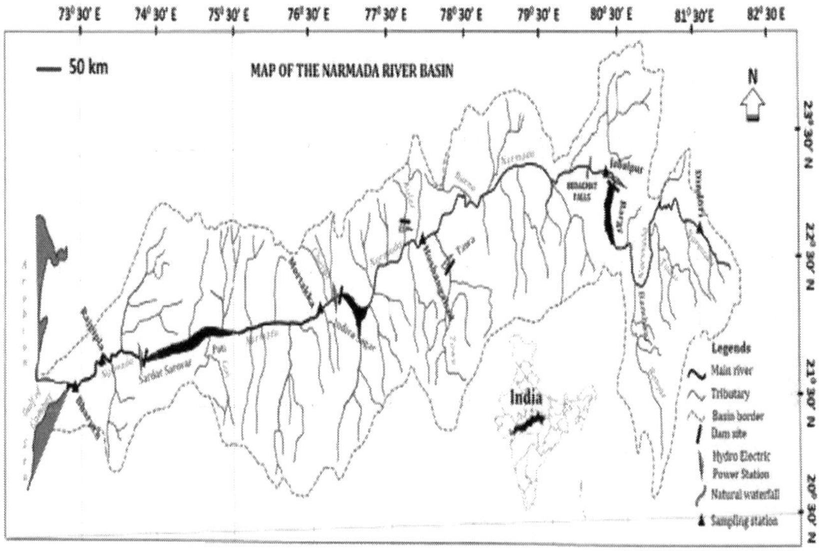

Figure 1: Map of the Narmada river basin showing location of sampling stations and dams.

Sampling

Fish were collected from six sampling stations using various nets and gears during April, 2011 - February, 2013. Sites were considered for their potential to accommodate cyprinids and mastacembelids (presence of lentic and lotic zones and vegetation shelters) and for their accessibility. Each sampling site was approximately 1500 m long, and all habitats were sampled downstream to upstream until at least 10-12 cyprinid and mastacembelid fish had been caught. We note that *C. catla* fish were not found at Dindori station, and likewise *M. armatus* were not found at Bharuch station. From each specimen a finclip, approximately 1 cm^2, was obtained and stored in absolute ethanol until laboratory analysis as describe below.

DNA Extraction, PCR and DNA Sequencing

Genomic DNA was extracted using the genomic DNA isolation kit (Promega wizard) from a total 43 *C. catla* and 60 *M. armatus* fish. Primers for amplification of the mitchondrial D-loop region of *M. armatus* were designed using the programs PRIMER 3 (Rozen and Skaletsky 2000) and Oligo Calc (Kibbe 2007) by Simgene using a reference sequence obtained from Genbank (EU380216.1). The primers designed for *M. armatus* were: MADF (5′-TTATATGCATTCATTCAGGTACA-3′) and MADR (5′-TAGGGCCCATTTTAACATCT-3′). For *C. catla* the Carp-Pro and Carp-Phe primers were as described by Thai et al. (2004). Amplifications were carried out using an initial denaturation step at 95°C for 2 min followed by 35 cycles of 94°C for 30 s, 57°C for 1 min and 72°C for 1 min along with a final extension at 72°C for 5 min. The amplified fragments were processed for cycle sequencing using the BigDye® Terminator v.3.1 Cycle Sequencing Kit (Applied Biosystems, Inc.) followed by cleanup and bi-directional sequencing using an ABI 3130 Genetic analyzer (Applied Biosystems, Inc.).

Data Analysis

Sequence Alignments and Topographic Analysis

Sequences were aligned using Codon code aligner v4.0.3 (CodonCode Corporation, Dedham, MA, USA). Within population diversity was estimated by computing haplotype diversity (H) and nucleotide diversity (π) indices using DnaSP v5.10 (Librado and Rozas 2009) and Arlequin v3.5.1.2 (Excoffier and Lischer 2010). Hierarchical relationships among the populations were analyzed using AMOVA and genetic variance was partitioned using Fst. The phylogenetic analysis included finding the best substitution model in Modeltest 2.1.1 (Darriba et al. 2012) using Akaike Information Criterion (AIC) and Baysian Information Criterion (BIC). The haplotype network was computed using Network v4.6.1.1 (Bandelt et al. 1999) where the haplotype pairwise differences were used to determine the number of mutational steps between haplotypes. A statistical test initially developed to analyze selective neutrality of mutations was implemented to test demographic expansion in recent years (de Jong et al. 2011; Ramos-Onsins and Rozas 2002). As shown by de Jong et al. (2011), these tests are designed to distinguish between neutrally evolving sequences in mutation drift equilibrium and sequences evolving under non-neutral processes including directional and balancing selection and demographic expansion or population contraction. These tests were performed in Arlequin v3.1 using 1000 simulations under a selective model of neutrality. For Mantel test, isolation by distance model was analyzed following the method of Jensen et al. (2005).

RESULTS

Diversity Indices

Mitochondrial DNA control region (D loop) sequences of 800 bp were obtained and aligned for the individuals from the Narmada river from to two fish species, *M. armatus* and *C. catla* (NCBI Accession numbers KF468051 to KF468110). The sequences of *M. armatus* (n = 60) showed 39 polymorphic sites, accounting for almost 4% of total number of sites examined. Among the polymorphic sites, 7 were singleton variable sites and 2 were parsimony informative sites. Similarly among the individuals representing the *C. catla* population (n = 42), there were 15 variable sites (~2%) including 6 singletons and one parsimony informative site.

Genetic Diversity

Eight haplotypes were found in the *M. armatus* populations and thirteen were found in the *C. catla*populations. The distribution and frequency of the haplotypes in each population is shown in Table 2. Haplotypes 2 and 3 are widely distributed in all populations of *M. armatus*. Five unique haplotypes, represented by only single individuals from Jabalpur, Mortakka and Rajpipla populations and three others were found in more than one population (Figure 2). In the *C. catla*populations, haplotype 2 is widely distributed whereas six unique haplotypes represented by only one population and seven others were admixtures of more than one population. Three were found to occur only in the Hoshangabad population (Table 2; Figure 3). Neighbor-joining (NJ) trees showing haplotype relationships were constructed based on the Kimura 2-parameter model, and bootstrap values based on 1,000 replicates shown (Figure 4A and B).

Table 2: Distribution and frequency of different D-loop haplotypes of *M. armatus*and *C. catla* in different populations

Haplotype	Dindori	Jabalpur	Hosangabad	Mortakka	Rajpipla	**Bharuch**
Mastacembelus armatus						
Hap1	4 *(0.40)*	0	0	1 *(0.076)*	0	--
Hap2	4 *(0.40)*	3 *(0.214)*	12 *(0.923)*	8 *(0.615)*	6 *(0.6)*	--
Hap3	2 *(0.20)*	4 *(0.286)*	1 *(0.076)*	3 (0.231)	3 *(0.3)*	--
Hap4	0	5 *(0.357)*	0	0	0	--
Hap5	0	1 (0.071)	0	0	0	--
Hap6	0	1 (0.071)	0	0	0	--
Hap7	0	0	0	1 *(0.076)*	0	--
Hap8	0	0	0	0	1 *(0.1)*	--
Catla catla						
Hap1	--	0	0	0	0	1 *(0.10)*
Hap2	--	0	1 *(0.125)*	2 *(0.25)*	2 *(0.20)*	4 *(0.40)*
Hap3	--	0	0	1 *(0.125)*	0	1 *(0.10)*
Hap4	--	0	3 *(0.375)*	0	0	1 *(0.10)*
Hap5	--	0	0	2 *(0.25)*	0	2 *(0.20)*
Hap6	--	4 *(0.667)*	0	0	1 *(0.10)*	1 *(0.10)*
Hap7	--	1 *(0.167)*	1 *(0.125)*	0	0	0
Hap8	--	0	1 *(0.125)*	0	0	0
Hap9	--	0	1 *(0.125)*	0	0	0
Hap10	--	0	1 *(0.125)*	0	0	0
Hap11	--	1 *(0.167)*	0	0	0	0
Hap12	--	0	0	3 *(0.375)*	6 *(0.60)*	0
Hap13	--	0	0	0	1 *(0.10)*	0

Italicized Values in bracket shows the relative haplotype frequency.

Khedkar *et al.*

Khedkar *et al. SpringerPlus* 2014 **3**:385, doi:10.1186/2193-1801-3-385

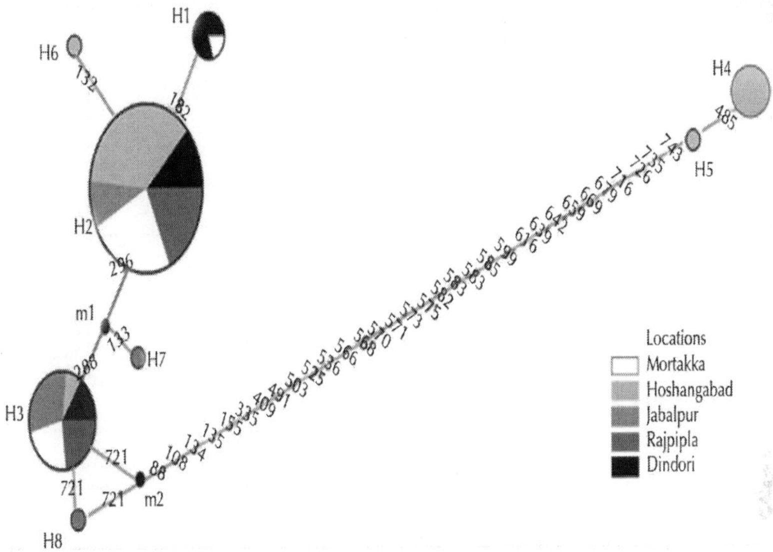

Figure 2: Median joining network map of haplotypes of *M. armatus*.

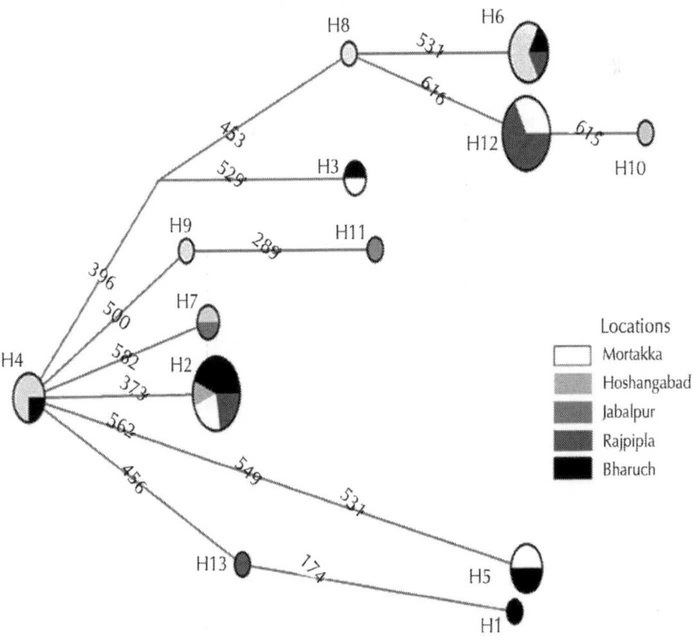

Figure 3: Median joining network map of haplotypes of *C. catla*.

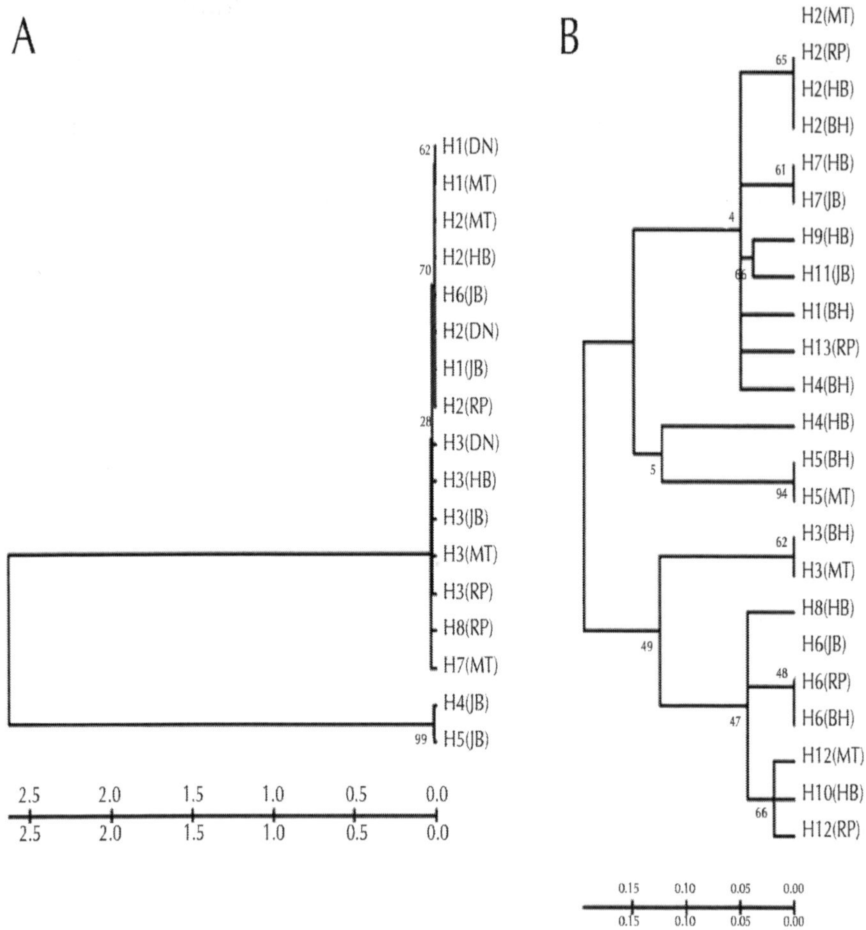

Figure 4: Haplotype based NJ tree (A) *M. armatus*; (B) *C. catla*.

For *M. armatus*, the haplotype diversity and nucleotide diversity values for the Jabalpur population were highest (0.791 and 0.0240 respectively) whereas these values were lowest (0.154 and 0.0003 respectively) in the Hoshangabad population (Table 3). In *C. catla,* for the population from Hoshangabad the haplotype diversity was also highest (0.8929), but for this species the nucleotide diversity was highest (0.2404) in the population from Mortakka (Table 3).

Table 3: D-loop sequence divergence values for *M. armatus* and *C. catla* for populations from the Narmada river

Fish populations	Parameters	Sampling stations					
		Dindori	**Jabalpur**	**Hoshangabad**	**Mortakka**	**Rajpipla**	**Bharuch**
Mastacembelus armatus	No. of Samples (n)	10	14	13	13	10	-
	No. of Polymorphic sites (PS)	3	39	2	4	3	-
	No. of Haplotypes (k)	3	5	2	4	3	-
	Haplotype diversity (H)	0.711 ± 0.0860	0.791 ± 0.0673	0.154 ± 0.1261	0.603 ± 0.1306	0.600 ± 0.1305	-
	Nucleotide diversity (π)	0.0015 ± 0.0011	0.0240 ± 0.0127	0.0003 ± 0.0004	0.0014 ± 0.0010	0.0015 ± 0.0011	-
Catla catla	No. of Samples (n)	--	6	8	8	10	10
	No. of Polymorphic sites (PS)	--	6	7	8	6	10
	No. of Haplotypes (k)	--	3	6	4	4	6
	Haplotype diversity (H)	--	0.6000 ± 0.2152	0.8929 ± 0.1113	0.8214 ± 0.1007	0.6444 ± 0.1518	0.8444 ± 0.1029
	Nucleotide diversity (π)	--	0.17333 ± 0.1241	0.1404 ± 0.0994	0.2404 ± 0.1551	0.1481 ± 0.1007	0.2044 ± 0.13120

Khedkar *et al.*

Khedkar *et al.* *SpringerPlus* 2014 **3**:385, doi:10.1186/2193-1801-3-385

Fst Analysis

Pairwise comparisons of the genetic variation contained in subpopulations relative to the total populations (Fst) of *M. armatus* were significant when the Jabalpur population was compared with all of the others sampled here. In addition, the comparison of the Dindori and Hoshangbad populations was also significant (Table 4). For the C. *catla* populations, these values are significant only for the comparisons of the Rajpipla and Bharuch populations and Jabalpur and Rajpipla populations (Table 5).

Table 4: Population pairwise Fst comparisons for *M. armatus* (below the diagonal)

	Dindori	Jabalpur	Hoshangabad	Mortakka	Rajpipla
Dindori	--	*0.00901**	*0.02703**	*0.45946*	*0.22523*
Jabalpur	0.33467	--	*0.00000**	*0.00000**	*0.03604**
Hoshangabad	0.15246	0.38583	--	*0.32432*	*0.08108*
Mortakka	0.01433	0.35912	0.04264	--	*0.62162*
Rajpipla	0.09018	0.31342	0.17758	-0.04100	--

*significant p values (<0.05).

Khedkar *et al.*

Khedkar *et al. SpringerPlus* 2014 **3**:385, doi:10.1186/2193-1801-3-385

Table 5: Population pairwise Fst comparisons for C. *catla (below the diagonal)*

	Jabalpur	Hoshangabad	Mortakka	Rajpipla	Bharuch
Jabalpur	--	*0.05405*	*0.18018*	*0.04505**	*0.06306*
Hoshangabad	0.18719	--	*0.38739*	*0.07207*	*0.29730*
Mortakka	0.07238	0.01690	--	*0.38739*	*0.34234*
Rajpipla	0.17569	0.17440	0.01864	--	*0.00901**
Bharuch	0.18310	0.02768	-0.00143	0.25298	--

*significant p values (<0.05).

Khedkar *et al.*

Khedkar *et al.* *SpringerPlus* 2014 **3**:385, doi:10.1186/2193-1801-3-385

For *M. armatus*, the low pairwise Fst value were noted (-0.0410) between Mortakka and Rajpipla may indicate that these populations undergo genetic exchange events. Among the *C. catla*populations, the low Fst values for the Mortakka and Bharuch comparison (-0.00143) may also suggest high levels of genetic exchange (Tables 4 and 5). Also, based on the AMOVA analysis, the majority of the variation was found to be within populations for both species (Table 6).

Table 6: Analysis of molecular variance (AMOVA) among and within populations

Source of variation	**Among populations of *M. armatus***	**Within *M. armatus*populations**	**Total**	**Fst**	**Among populations of C. *catla***	**Within C. *catla*populations**	**Total**	**Fst**
d.f.	4	55	59	0.3833*	4	37	41	0.114**
Sum of squares	76.321	148.212	224.533		11.295	50.300	61.595	
Variance component	1.372Va	2.694Vb	4.066		0.175Va	1.359Vb	1.359	
Percentage of variation	33.74	66.26	--		11.45	88.55	--	

Va-Variation among groups; Vb-Variation among populations within groups.

*significant p values (<0.05), **significant p values (<0.01).

Khedkar *et al.*

Khedkar *et al. SpringerPlus* 2014 **3**:385 doi:10.1186/2193-1801-3-385

To further analyze the potential for gene flow among populations at different locations, we conducted an analysis in ARLEQUIN to obtain M values (Tables 7 and 8). A value greater than two here indicates gene flow between populations (Mallet 2001; Hebert et al. 2004). More than half of our M values are greater than two, and for each of the species, one of the M values is infinite.

Table 7: Matrix of M values for *M. armatus*

	Dindori	**Jabalpur**	**Hoshangabad**	**Mortakka**	**Rajpipla**
Dindori	--				
Jabalpur	0.99402	--			
Hoshangabad	2.77964	0.79590	--		
Mortakka	34.40182	0.89230	11.22727	--	
Rajpipla	5.04464	1.09529	2.31561	infinite	--

Khedkar *et al.*

Khedkar *et al. SpringerPlus* 2014 **3**:385, doi:10.1186/2193-1801-3-385

Table 8: Matrix of M values for *C. catla*

	Jabalpur	**Hoshangabad**	**Mortakka**	**Rajpipla**	**Bharuch**
Jabalpur	--				
Hoshangabad	2.17115	--			
Mortakka	6.40838	29.09091	--		
Rajpipla	2.34597	2.36696	26.32727	--	
Bharuch	2.23077	17.56350	infinite	1.47643	--

Khedkar *et al.*

Khedkar *et al. SpringerPlus* 2014 **3**:385, doi:10.1186/2193-1801-3-385

Non-significant correlations for isolation by distance were noticed in *M. armatus* populations (Mantel Test; $Z = 472.20$, $r = -0.0322$, one sided $P = 0.868$ from 10,000 randomizations). RMA regression analysis also revealed a negative isolation-by-distance relationship in case of *M. armatus* ($y = 0.2607x - 2.837$, $r^2 = 0.104$) (Figure 5A). For the *C. catla* populations there was a positive correlation for isolation by distance

(Mantel test; Z=466.05, r=0.408, one sided P=0.1312 from 10000 randomizations) (Figure 5B).

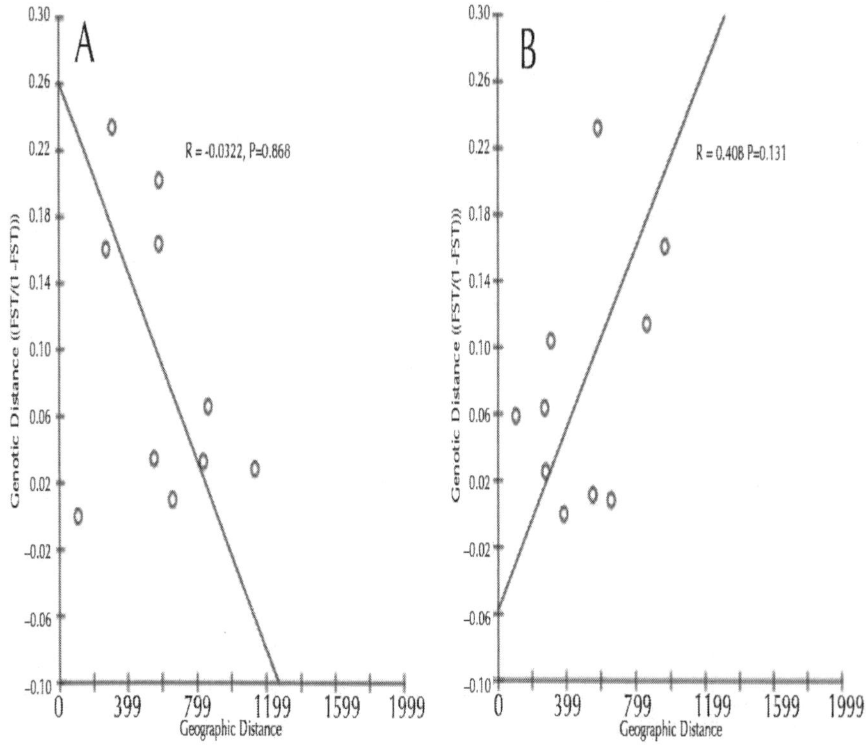

Figure 5: Scatter plot of genetic vs. geographic variation with RMA regression (A) *M. armatus;* (B) *C. catla*.

There was significant support for the model of sudden expansion of population size as well as spatial expansion for the *M. armatus* population at Jabalpur (SSD=0.7924,' p=0.000 and SSD=0.129, p=0.0400 respectively). In *C. catla* there is no significant support for either sudden expansion or spatial expansion in any of the populations. The values for Harpending's raggedness index were non-significant in all populations belonging to both species (Table 9).

Table 9: Analysis of populations to deduce demographic and spatial expansion events

Discussion

Statistics	Dindori	Jabalpur	Hoshangabad	Mortakka	Rajpipla	Bharuch
M. armatus Demographic expansion						
SSD*	0.05702	0.79242	0.03195	0.10070	0.08515	--
p- values	0.08000	0.00000	0.11000	0.10000	0.21000	--
Raggedness index	0.06765	0.37640	0.76331	0.36736	0.31111	--
p- values	0.77000	0.95000	0.66000	0.10000	0.24000	--
Spatial expansion						
SSD*	0.00738	0.12927	0.01297	0.07936	0.05586	--
p- values	0.61000	0.04000	0.20000	0.14000	0.21000	--
C. catla Demographic expansion						
SSD*	--	0.14614	0.01264	0.08586	0.19467	0.08154
p-values	--	0.10000	0.70000	0.04000	0.06000	0.07000
Raggedness index	--	0.27556	0.08291	0.24362	0.62519	0.19605
p- values	--	0.33000	0.73000	0.12000	0.05000	0.08000
Spatial expansion						
SSD*	--	0.05272	0.01264	0.06637	0.13729	0.07079
p- values	--	0.50000	0.67000	0.34000	0.09000	0.27000

(*SSD-sum of squared deviations).

Khedkar *et al.*

Khedkar *et al. SpringerPlus* 2014 **3**:385, doi:10.1186/2193-1801-3-385

DISCUSSION

Concerns over habitat alterations and the subsequent effects on biodiversity have received much scientific attention during last few decades. Particularly for aquatic diversity, conservationists are concerned that dams may fragment and diminish natural habitats in rivers (Dynesius and Nilsson1994; Khedkar et al. 2014a) and create environmental disparities (Ward and Stanford 1983) that can result in the loss of genetic diversity, reduced population sizes and inbreeding

(Ellstrand and Elam 1993; Lynch et al.1995). Ultimately, this may lead to population extinctions (Fagan 2002). Evaluation of Indian rivers for the possible effects of fragmentation of river habitats has not been adequately addressed, and in part this study is first effort of its kind to evaluate the possible effects of fragmentation of the Narmada river of India using genetic approaches. Prior to this study, no records of genetic studies for the Narmada river fish have been found (Khedkar et al.2014b).

The results of our study indicate several things. First, using the data from mtDNA control region (D loop), overall levels of genetic variation of *M. armatus* are moderate as compared to other teleostean fishes (Panarari-Antunes et al. 2012; Librado and Rozas 2009; Aboim et al. 2005; Stefanni and Thorley 2003; Salzburger et al. 2003; Lee et al. 1995; Fajen and Breden 1992). In terms of populations, the *M. armatus* population from Jabalpur is naturally isolated due to a waterfall fall on downstream side and a dam on the upstream side. This population had high levels of nucleotide diversity (0.0240), with up to 39 nucleotide substitutions in some cases. This suggests that this population is evolving independently and may become further genetically differentiated. Similar observations were made for populations of *P. squamosissimus* species from the Parniba and Tocantinus river basins in China (Panarari-Antunes et al. 2012).

In contrast, genetic differentiation among the *C. catla* populations in the Narmada river basin appears to be weaker as overall levels of genetic diversity were only moderate to high. Most of the dams constructed across the river basin are almost 30 years old, and the Bedaghat fall is a natural barrier that is much older. The MtDNA data described here uphold the common assumption that analyzed populations of two native fish species existed in this river before the construction of the dams.

From samples across the entire river, we found eight haplotypes among 60 individuals of *M. armatus* vs. thirteen haplotypes among 42 individuals of *C. catla*. The range of haplotype diversity values (H = 0.154 to 0.791 and 0.6000 to 0.8929, respectively for these two species) and nucleotide diversity values ($\pi = 0.0003$ to 0.0240 and 0.1404 to 0.2044 respectively) (Table 3) were similar to values seen in studies by Chen et al. (2006) on Kessle fishes (H = 0.9992, $\pi = 0.0082$), Anthunes et al. (2012) on *P. squamonsissmus* (H = 0.690, $\pi = 0.0236$).

Genetic data presented here suggest the possibility that *M. armatus* populations, which are invasive to the Narmada basin, originated from Jabalpur. This region was naturally isolated from the rest of downstream regions by the Bedaghat falls a few million years ago, and this would naturally limit genetic exchange for this population on the upstream side. The Jabalpur population has relatively moderate genetic and haplotype diversity, and this may also suggest a recent population expansion after a founder event (Shaw et al. 1992; Carvalho et al. 1996). Comparisons of populations at localities serving as drainages, however, could also provide additional resolution for assessing the levels of genetic differentiation of *M. armatus* populations.

The data revealed different levels of genetic differentiation among the different sampling locations across the basin of the Narmada, Haplotypes 2 and 3 may be treated as ancient haplotypes because they are distributed among all populations, whereas the few private haplotypes may be more recently evolved (Figure 2). The genetic diversity within *C. catla* populations documented here strongly suggests that these fishes were likely derived from different sources, possibly through human translocations, since a mixed haplotype distribution was observed (Figure 3). No significant genetic divergences were revealed in *C. catla*. This may be explained by high rates of gene flow between populations of this species within this system.

C. catla is also widely used for aquaculture in India because of its simple culture needs and defined seed production technologies (Jhingran 1968). Many hatcheries are located in the Narmada basin, particularly at the Bharuch region, and artificial seed products may be reentering the river channel further impacting genetic diversity. Also from the median joining networks, a higher degree of haplotype sharing between localities was seen for *M. armatus* as compared to the *C. catla*populations (Figures 2 and 3). Therefore, the patterns of genetic diversity seen within *C. catla* may also reflect both inbreeding and genetic drift type events (Nei et al. 1975; Wishard et al. 1984).

For *M. armatus*, the pairwise Fst values indicate some significant amounts of genetic exchange between the Mortakka and Rajpipla populations in spite of the artificial barrier created by the Sardar Sarowar dam. This suggests that these locations may be connected through a sub tributary network that allows migration. On the other hand, the significantly lower amounts of genetic exchange between the Jabalpur and Dindori populations supports the idea that the presence of the Barghi dam between them promotes fragmentation. Also overall, the pairwise Fst values for the Jabalpur population of *M. armatus* compared to the other populations also indicates significant isolation and supports the role of various barriers in promoting population isolation (Table 4). In the case of *C. catla*, the Fst pairwise comparisons suggest that between some populations, such as the Bharuch and Rajpipla populations, significant amounts of genetic exchange can occur. Here, there is no barrier separating these locations.

The potential for isolation is also considered in the analysis of M values. Here, values of M below 2 imply isolation in the population structure (Mallet 2001). The pairwise comparisons of M values for the *M. armatus* population from Jabalpur with Dindori, Hoshangabad, Mortakka and Rajpipla are all less than this critical value, and this supports the role of dams in promoting genetic fragmentation and population structuring (Table 7). The populations of *C. catla* do not show such structuring, except for the pairwise comparison of M values between the Bharuch and Rajpipla population ($M<2$). This result is consistent with the results based on pairwise Fst values as described in the previous section.

The results of demographic analysis by mismatch distribution show evidence of population expansion. This inference is supported, except perhaps for the Jabalpur population of *M. armatus*(Table 9). We can assume that a recent population expansion might have been affected by genetic drift due to a bottleneck or a founder effect in case of *M. armatus*. The demographic and spatial expansion values for *M. armatus* are significant for Jabalpur population only (Table 9).

Isolation by Distance

Our sampling was restricted to the main channel of the Narmada River, and the sites are lined up in accordance with the stepping stone model of Kimura (1953). According to this model (Kimura and Weiss 1964), in the idealized case the genetic distance between samples increases simply as geographical distance increases. This model also predicts that in the one-dimensional case, the genetic correlation should fall off exponentially with distance. In the case of *M. armatus*, our observations are not in accordance with this predicted pattern because here adjacent populations are significantly differentiated. This is consistent with the notion that other factors, such as the barriers separating these populations, are promoting fragmentation. In contrast, the relationships of the *C. catla* population follow the model pattern in general and do not appear to be so fragmented.

Upstream-Downstream Structure

The variation in genetic diversity along the upstream-downstream gradient is another parameter potentially influenced by the presence of dams. We observed that haplotype and nucleotide diversities are not lined up in a manner consistent with a simple upstream-downstream gradient in populations of both fish species. However, haplotype diversity may be sensitive to sampling bias in that only about 10 individuals per population collected at each sampling site. Additional sampling may reveal the presence of undetected haplotypes.

A number of factors may explain the genetic diversity values seen for the *M. armatus* populations. First, the upper part of the Narmada river studied here is characterized by a natural waterfall and a narrow channel, and it is known that fish population sizes are often correlated to the amount of available habitat (Frankham 1996; Hanfling et al. 2002). Also, *M. armatus* may be subject to a range-edge effect (Arnaud-Haond et al. 2006) since it was more difficult to catch enough specimens in the sites furthest upstream. No individuals were caught at the last locality (Bharuch). Specifically for *M. armatus*, the nature of the dispersion between sampling points, due to the river flow and the obstacles, also suggests that the downstream sites are not receiving new haplotypes from upstream localities.

In the case of *C. catla*, this same dispersion process may maintain a high level of genetic diversity downstream, while genetic drift is less influential in the upstream sampling sites (Lacy 1987). Combining these observations, it can be hypothesized that the *M. armatus* populations within the Narmada river are organized as a single metapopulation composed of local populations exchanging individuals by migration processes (Levins 1970; Hanski and Gilpin 1991). Therefore in this case, the applicability of the stepping-stone model can be rejected.

Dams and or waterfalls, which are designed to prevent upstream movements, seem here in some cases to be sufficient obstacles to induce detectable genetic consequences. In other studies of small species such as bullhead, it has been shown that obstacles only about 20 cm high can prevent upstream movement. Individuals of species such as *Cottus gobio* (Utzinger et al. 1998) and barbel (*Barbus barbus*) individuals are stopped or slowed in their upstream dispersion by a barrier 40 cm high weir (Lucas and Frear 1997). This study confirms that in some cases both dams and water falls can contribute to the asymmetry of dispersion and a decrease in genetic diversity of fish populations.

Data Set Quality and Potential Bias

The Sampling Scheme

Our sampling strategy was to survey as many sites as possible between obstacles that may impact the river habitat structure. The main objective of this study was to assess the extent to which genetic differentiation of populations can be related to fragmentation. The Narmada river system was chosen because there has been extensive fragmentation due to the presence of waterfalls and dams.

In artificially fragmented systems, the number of obstacles is strongly correlated with water way distances (Meldgaard et al. 2003). Dams are generally positioned such that the distance between two consecutive obstacles is about 250 km on average. This holds true for the Narmada system, except that there is also a natural waterfall within 60 kms of the Barghi dam. Due to this spatial configuration, here the effects of waterway distance and the number of obstacles were studied independently.

Lastly, only sampling points of the main Narmada River were considered here. Tributaries are known to have influences on communities inhabiting the main stream of drainage basins, and could also be involved in the demographic and genetic structuring of the Narmada populations (Hitt and Angermeier 2006; Khedkar et al. 2014a). However, for some species, Carlsson et al. (1999) suggested that tributaries may represent ecologically differentiated and locally adapted populations that only rarely disperse in to the main stream, and vice versa. Thus, consideration of only the main Narmada river may be an efficient way to study the genetic structure of these fish populations, although it would be useful to investigate in future studies. the role potentially played by the tributaries.

CONCLUSIONS

The results reported herein for *M. armatus* and *C. catla* populations from the Narmada river clearly show evidence, in some cases, for fragmentation effects by dams and a water fall, and that these barriers contribute to the genetic isolation and differentiation of fish populations. The prevention of movements, either completely from downstream to upstream, or partially from upstream to downstream, by dams or other barriers, enhance the natural effects of isolation by distance and the asymmetry of the dispersal flows. Consequently, populations, especially those furthest upstream, would be expected to have very low immigration rates and be more subject to genetic impoverishment. This study can also provide methodological guidance for future studies of such ecological situations. Sampling a long river segment such as the Narmada with multiple obstacles is certainly a more efficient way to assess the impact on fish dispersal.

Finally, our study does suffer from some biases. Some sample sizes were low, and in some cases it was not possible directly sample upstream and downstream of each barrier. Additional studies in the future may address these biases by including samples of tributaries and other potential sources of input into these populations.

AUTHOR' CONTRIBUTIONS

Conceived and designed the experiments: GD TBR DH. Performed the experiments: GD RJ AK AT. Analyzed the data: GD DH. Contributed reagents/materials/analysis tools: GD. Wrote the paper: GD TBR DH. All authors read and approved the final manuscript.

ACKNOWLEDGEMENTS

Funding from NAIP, ICAR, New Delhi, Ministry of Food processing Industries, New Delhi, University Grants commission, New Delhi and CREST Fellowship by Department of Biotechnology, Government of India to GD Khedkar is deeply acknowledged. We also thank to the authorities of Paul Hebert Centre for DNA Barcoding and Biodiversity Studies for providing access to the molecular genetic analysis facilities.

REFERENCES

1. Aboim MA, Menezes GM, Schlitt T, Rogers AD (2005) Genetic structure and history of populations of the deep-sea fish *Helicolenus dactylopterus* (Delaroche, 1809) inferred form mtDNA sequence analysis. Mol Ecol 14:1343-1354
2. Antunes A, Faria R, Johnson WE, Guyomard RP, Alexandrino P (2006) Life on the edge: the long-term persistence and contrasting spatial genetic structure of distinct brown trout life histories at their ecological limits. J Heredity 97:193-205
3. Arnaud-Haond S, Teixeira S, Massa SI, Billot C, Saenger P, Coupland G, Duarte CM, Serrao EA (2006) Genetic structure at range edge: low diversity and high inbreeding in Southeast Asian mangrove (*Avicennia marina*) populations. Mol Ecol 15:3515-3525
4. Bainbridge R (1958) The speed of swimming of fish as related to size and to the frequency and amplitude of the tail beat. J Exper Biol 35:109-133
5. Bandelt HJ, Forster P, Röhl A (1999) Median-joining networks for inferring intraspecific phylogenies. Mol Biol Evol 16:37-48
6. Bremer JRA, Mejuto J, Gomez-Marquez J, Boan F, Carpintero P, Rodrıguez JM, Vinas J, Greig TW, Ely B (1996) Hierarchical analyses of genetic variation of samples from breeding and feeding grounds confirm the genetic partitioning of northwest Atlantic and South Atlantic populations of swordfish (*Xiphias gladius* L.). J Exp Mar Biol Ecol 327:167-182

7. Carlsson J, Olsen KH, Nilsson J, Overli O, Stabell OB (1999) Microsatellites reveal fine-scale genetic structure in stream-living brown trout. J Fish Biol 55:1290523303
8. Carvalho GR, Shaw PW, Hauser L, Seghers BH, Magurran AE (1996) Artificial introductions, evolutionary change and population differentiation in Trinidadian guppies (*Poecilia reticulata*: Poeciliidae). Biol J Linn Soc 57:219-234
9. Chen D, Zhang C, Lu C, Zhang X (2006) Polymorphism of D-loop sequence from mitochondrial genomes of different brood-stocks of *Gymnocypris przewalskii* (Kessler). J Fishery Sci China 13:800-805
10. CodonCode Corporation 101 Victoria Street Centerville, MA 02632. http://www.codoncode.com
11. Darriba D, Taboada GL, Doallo R, Posada D (2012) jModelTest 2: more models, new heuristics and parallel computing. Nat Methods 9(8):772
12. de Jong MA, Wahlberg N, van Eijk M, Brakefield PM, Zwaan BJ (2011) Mitochondrial DNA signature for range-wide populations of *Bicyclus anynana* suggests a rapid expansion from recent refugia. PLoS One 6(6):e21385 doi:10.1371/journal.pone.0021385
13. Dunham JB, Vinyard GL, Rieman BE (1997) Habitat fragmentation and extinction risk of Lahontan cutthroat trout. N A J Fish Man 17:1126-1133
14. Dynesius M, Nilsson C (1994) Fragmentation and flow regulation of river systems in the northern third of the world. Sci (Washington, DC) 266:753-762
15. Ellstrand NC, Elam DR (1993) Population genetic consequences of small population size: implications for plant conservation. A Rev Ecol Syst 24:217-242
16. Excoffier L, Lischer HEL (2010) Arlequin suite ver 3.5: a new series of programs to perform population genetics analyses under Linux and Windows. Mol Ecol Reso 10:564-567
17. Fagan WF (2002) Connectivity, fragmentation, and extinction risk in dendritic meta populations. Ecology 83:3243-3249
18. Fajen A, Breden F (1992) Mitochondrial DNA sequence variation among natural populations of the Trinidad guppy, *Poecilia reticulata*. Evolution 46:1457-1465

19. Frankham R (1996) Relationship of genetic variation to population size in wildlife. Con Biol 10:1500-1508
20. Froese R, Binohlan C (2000) Empirical relationships to estimate asymptotic length, length at first maturity and length at maximum yield per recruit in fishes, with a simple method to evaluate length frequency data. J Fish Biol 56(4):758-773
21. Hanfling B, Hellemans B, Volckaert FAM, Carvalho GR (2002) Late glacial history of the cold-adapted freshwater fish *Cottus gobio*, revealed by microsatellites. Mol Ecol 11:1717-1729
22. Hanski I, Gilpin M (1991) Metapopulation dynamics: brief history and conceptual domain. Biol J Linn Soc 42:3-16
23. Hebert PDN, Penton EH, Burns JM, Janzen DH, Hallwachs W (2004) Ten species in one: DNA barcoding reveals cryptic species in the neotropical skipper butterfly *Astraptes fulgerator*. Proc Natl Acad Sci, USA 101:14812-14817
24. Hitt NP, Angermeier PL (2006) Effects of adjacent streams on local fish assemblage structure in western Virginia: implications for biomonitoring. A Fish Soc Symp 48:75-86
25. Jensen JL, Bohonak AJ, Kelley ST (2005) Isolation by distance, web service. BMC Gen 6:13 v.3.23 http://ibdws.sdsu.edu
26. Jhingran VG (1968) Synopsis of biological data on ostia, Catla catia (Hamilton, 1822)14-6F290 FAO Fich Synnpc, (32) Rev.1:pag.var.
27. Khedkar GD, Chandrashekar Reddy A, Ron TB, Haymer D (2013) High levels of genetic diversity in*Penaeus monodon* populations from the east coast of India. SpringerPlus 2:671
28. Khedkar GD, Lutzky S, Rathod S, Kalyankar A, David L (2014a) A dual role of dams in fragmentation and support of fish diversity across the Godavari River basin in India. Ecohydrol. doi:10.1002/eco.1470
29. Khedkar GD, Jamdade R, Naik S, David L, Haymer D (2014b) DNA Barcodes for the FIshes of the Narmada, One of India's Longest Rivers. PLoS ONE 9(7):e101460 doi:10.1371/journal.pone.0101460
30. Kibbe WA (2007) OligoCalc: an online oligonucleotide properties calculator. N Acids Res 35(suppl 2):W43-W46

31. Kimura M (1953) Stepping-stone model of population. Annu Rep Natl Inst Genet 3:62-63

32. Kimura M, Weiss GH (1964) The stepping stone model of population structure and the decrease of genetic correlation with distance. Genetics 49:561-576

33. Knaepkens G, Verheyen E, Galbusera P, Eens M (2004) The use of genetic tools for the evaluation of a potential migration barrier for the bullhead. J Fish Biol 64:1737-1744

34. Lacy RC (1987) Loss of genetic diversity from managed populations: interacting effects of drift, mutation, immigration, selection, and population subdivision. Conser Biol 1:143-158

35. Lee WJ, Conroy J, Howell WH, Kocher TD (1995) Structure and evolution of teleost mitochondrial control regions. J Mol Evol 41:54-66

36. Levins R (1970) Extinction. In: Gustenhaver M (ed) Some mathematical problems in biology, American Mathematical Society, Providence, Rhode Island. pp 76-107

37. Librado P, Rozas J (2009) DnaSP v5: A software for comprehensive analysis of DNA polymorphism data. Bioinformatics 25:1451-1452

38. Lucas MC, Frear PA (1997) Effects of a flow-gauging weir on the migratory behavior of adult barbel, a riverine cyprinid. J Fish Biol 50:382-396

39. Lynch M, Conery J, Burger R (1995) Mutation accumulation and the extinction of small populations. Am Nat 146:489-518

40. Mallet J (2001) Gene Flow. In: Woiwod IP, Reynolds DR, Thomas CD (eds) Insect Movement: Mechanisms and Consequences, CAB International, Wallingford, UK. pp 337-360

41. McKeown BA (1984) Fish Migration. Croom Helm Ltd., London.

42. Meldgaard T, Nielsen EE, Loeschcke V (2003) Fragmentation by weirs in a riverine system: a study of genetic variation in time and space among populations of European grayling (Thymallus thymallus) in a Danish river system. Con Gene 4:735-747

43. Miranda R, Oscoz J, Leunda PM, Garcia-Fresca C, Escala MC (2005) Effects of weir construction on fish population structure in the River Erro (North of Spain). Annales De Limnologie-Int J Limn 41:7-13

44. Morita K, Yamamoto S (2002) Effects of habitat fragmentation by damming on the persistence of stream-dwelling charr populations. Cons Biol 16:1318-1323
45. Nei M, Maruyama T, Chakraborty R (1975) The bottleneck effect and genetic variability in populations. Evolution 29:1-10
46. Neraas LP, Spruell P (2001) Fragmentation of riverine systems: the genetic effects of dams on bull trout (*Salvelinus confluentus*) in the Clark Fork River system. Mol Ecol 10:1153-1164
47. Northcote TG (1984) Mechanisms Of Fish Migration In Rivers. In: McCleave JD, Arnold GP, Dodson JJ, Neill WH (eds) Mechanisms Of Migration In Fishes, Plenum, New-York. pp 317-355
48. Nyakaana S, Arctander P, Siegismund HR (2002) Population structure of the African savannah elephant inferred from mitochondrial control region sequences and nuclear microsatellite loci. Heredity 89:90-98
49. Ovidio M, Philippart JC (2002) The impact of small physical obstacles on upstream movements of six species of fish -Synthesis of a 5-year telemetry study in the River Meuse basin. Hydrobiologia 483:55-69
50. Panarari-Antunes RS, Prioli AJ, Prioli SMAP, Júlio HF, Agostinho CS, Prioli LM (2012) Molecular variability in Brycon cf. pesu Müller and Troschel, 1845 (Characiformes: Characidae) from the Araguaia-Tocantins Basin. Gen Mole Res: GMR 7(1):95-106 DOI:10.4238/vol7-1gmr365
51. Poulet N (2007) Impact of weirs on fish communities in a piedmont stream. River Res Appli 23:1038-1047
52. Rainboth WJ (1996) Fishes of the Cambodian Mekong. FAO Species Identification Field Guide for Fishery Purposes. FAO, Rome. p 265
53. Ramos-Onsins SE, Rozas J (2002) Statistical properties of new neutrality tests against population growth. Mol Biol Evol 19:2092-2100 doi:10.1093/oxfordjournals.molbev.a004034
54. Rao KL (1975) India's Water Wealth. Orient Longman Ltd., New Delhi. p 255
55. Rozen S, Skaletsky H (2000) Primer3 on the WWW for General Users And For Biologist Programmers. In: Krawetz S, Misener

S (eds) Bioinformatics Methods And Protocols: Methods In Molecular Biology, Humana Press, Totowa, NJ. pp 365-386

56. Salzburger W, Brandstatter A, Gilles A, Parson W, Hempel M, Sturmbauer C, Meyer A (2003) Phylogeography of the vairone (*Leuciscus souffia,* Risso 1826) in Central Europe. Mol Ecol 12:2371-2386
57. Sato S, Kojima H, Ando J, Ando H, Wilmot RL, Seeb LW, Efremov V, LeClair L, Buchholz W, Jin DH, Urawa S, Kaeriyama M, Urano A, Abe S (2004) Genetic population structure of chum salmon in the Pacific Rim inferred from mitochondrial DNA sequence variation. Env Biol Fishes 69:37-50
58. Shaw PW, Carvalho GR, Seghers BH, Magurran AE (1992) Genetic consequences of an artificial introduction of guppies (*Poecilia reticulata*) in N. Trinidad. Proc R Soc B-Biol Sci 248:111-116
59. Stefanni S, Thorley JL (2003) Mitochondrial DNA phylogeography reveals the existence of an evolutionary significant unit of the sand goby *Pomatoschistus minutus* in the Adriatic (Eastern Mediterranean). Mol Phyl Evol 28:601-609
60. Thai BT, Burridge CP, Pham TA, Austin CM (2004) Using mitochondrial nucleotide sequences to investigate diversity and genealogical relationships within common carp (*Cyprinus carpio* L.). Ani Gen 36:23-28
61. Utzinger J, Roth C, Peter A (1998) Effects of environmental parameters on the distribution of bullhead *Cottus gobio* with particular consideration of the effects of obstructions. J App Ecol 35:882-892
62. Vidthayanon C (2002) Peat Swamp Fishes of Thailand. Office of Environmental Policy and Planning, Bangkok, Thailand. p 136
63. Ward JV, Stanford JA (1983) The Serial Discontinuity Concept Of Lotic Ecosystems. In: Fontaine TD, Bartell SM (eds) Dynamics Of Lotic Ecosystems, Ann Arbor Science, Ann Arbour, MI, USA. p 494
64. Wilson AC, Cann RL (1985) Mitochondrial DNA and two perspectives on evolutionary genetics. Biol J Linn Soc 26:375-400
65. Winfield IJ, Townsend CR (1991) The Role Of Cyprinids In Ecosystems. In: Winfield IJ, Nelson JS (eds) Cyprinid Fishes:

Systematics, Biology And Exploitation, Chapman and Hall, London. pp 552-567

66. Wishard LN, Seeb JE, Utter FM, Stefan D (1984) A genetic investigation of suspected red band trout populations. Copeia 1:120-132

Chapter 6

Warming and Neighbor Removal Affect White Spruce Seedling Growth Differently Above and Below Treeline

Kyoko Okano and M Syndonia Bret-Harte

Institute of Arctic Biology, University of Alaska, Fairbanks, Fairbanks, AK, USA

ABSTRACT

Climate change is expected to be pronounced towards higher latitudes and altitudes. Warming triggers treeline and vegetation shifts, which may aggravate interspecific competition and affect biodiversity. This research tested the effects of a warming climate, habitat type, and neighboring plant competition on the establishment and growth of white spruce (*Picea glauca* (Moench) Voss) seedlings in a subarctic mountain region. *P. glauca* seedlings were planted in June 2010 under 4 different treatments (high/control temperatures, with/without competition) in 3 habitats (alpine ridge above treeline/tundra near treeline /forest below treeline habitats). After two growing seasons in 2011, growth, photosynthesis and foliar C and N data were obtained from a total of 156, one-and-a-half year old seedlings that had survived. Elevated temperatures increased growth and photosynthetic rates above and near treeline, but decreased them below treeline. Competition was increased by elevated temperatures in all habitat types. Our results suggest that increasing temperatures will have positive effects on the growth of *P. glauca* seedlings at the locations where *P. glauca* is expected to expand its habitat, but increasing temperatures may have negative effects on seedlings growing in mature forests. Due to interspecific competition, possibly belowground competition, the upslope expansion of treelines may not be as fast in the future as it was the last fifty years.

INTRODUCTION

Annual average temperature in Alaska has increased 1.9°C over the past 50 years and is projected to rise from 2.8°C to 7.2°C above the 1960s-1970s baseline by the end of this century (Karl et al. [2009]). The projection is higher than the estimated 2.4°C to 6.4°C rise in global average temperature over the same period of time (IPCC [2008]). This recent climate warming is expected to alter the relative abundance of arctic and subarctic vegetation in Alaska. Photographic comparisons of past and current vegetation (Sturm et al. [2001], Tape et al. [2006], Hamm [2007], Stueve et al. [2011]) have shown both northward and altitudinal expansions of shrubs and treeline into tundra. Studies in Denali National Park and Preserve (DNPP) reported the maximum treeline advance was 150 m upslope into tundra (Stueve et al. [2011]) and the average treeline advance was 40 to 60 m (Hamm [2007]) in the last 50 years. Those studies suggest that higher temperatures stimulate successful recruitment, and possibly growth of treeline species, predominantly *Picea glauca* (white spruce) in the subarctic mountain region.

P. glauca is also one of the dominant species in boreal forests; however, higher temperatures that have been observed more frequently since 1950 are linked to drought stress, resulting in decreased growth rates of *P. glauca* (Wilmking et al. [2004]). Because of desiccation, mature trees could become more susceptible to insect outbreaks, diseases and wildfires, thus increasing mortality. Models project that the growth rate of*P. glauca* could decrease to 0 - 20% of the long-term mean by 2100 due to drought stress, which might lead to species elimination (Juday et al. [2005]). Recent studies of marginal tree population have shown both positive and negative responses to climate warming (e.g. Wilmking et al. [2004]). Experimentation is necessary to uncover the mechanistic processes at work.

In addition to altering the growth of adult trees, climate change may have strong effects on sexual reproduction and the establishment of seedlings, which are the most vulnerable life stages of a tree. Warm temperatures are thought to be important for the establishment of seedlings and their survival during the first 50 years of growth (Szeicz and MacDonald [1995]). However, rather than temperature effect itself, the growth of *P. glauca* seedlings in subarctic alpine forests on south-facing slopes was limited by winter drought and cold-induced photoinhibition in the early spring and/or the early morning in the summer (Danby and Hik [2007], Slot et al. [2005]). Further, even if warming positively affects seedlings, drought stress could also be increased. Also, the viability of seeds from trees growing at higher elevations is usually lower than from those growing at lower elevations (Roland et al. [2013]), which may inhibit upslope expansion.

Seedlings of *P. glauca* at a higher elevation need to acclimate to harsh new abiotic conditions, and interact with other plant species that are already present. Positive interactions with neighboring plants are crucial for small spruce seedlings in harsh environments. Choler et al. ([2001]) found that the effects of neighboring plants are facilitative rather than competitive in high and exposed sites. However, Bret-Harte et al. ([2008]) reported that some arctic species responded positively to neighbor removal while others did not. The response of *P. glauca* seedlings to neighboring plants under a changing climate is thus, likely to be determined by complex factors that are not fully understood.

The goal of this study was to understand the mechanistic effects of warming and interspecific competition on the performance of seedlings of *P. glauca* in a range of habitats above and below the current treeline on subarctic Alaskan mountains, in order to inform future conservation. Our hypotheses are as follows. 1) Growth of *P. glauca* seedlings above treeline is reduced relative to seedlings below treeline due to negative environmental conditions such as low temperatures and low water content in soil. 2) Warming in a cold environment positively affects seedlings as long as no drought stress occurs. 3) Interactions between *P. glauca* and neighboring plants are mostly competitive, but harsh conditions can trigger less competitive or even facilitative effects. We planted *P. glauca* seedlings at three different habitats to test habitat effects (hypothesis 1). At each site, the seedlings were planted inside or outside a small greenhouse to compare the effects of warming vs. ambient temperatures (hypothesis 2). Both inside and outside the greenhouses, the seedlings were planted either with neighboring plants or with neighbors removed, to examine the interspecific interaction at ambient and raised temperatures (hypothesis 3). Our results will provide insight how seedlings of treeline species will respond to climate change, and how habitat type and interspecific competition affect treeline shifts and the establishment and growth of spruce seedlings.

RESULTS

Environmental Manipulation

Mean ambient air temperatures differed by 2°C - 3.5°C among three habitat types during the summer growing season in 2010 (Table 1). Because of a failure of a logger below treeline, air temperature data in 2011 were obtained only from the above and near treeline sites, whose difference was only 1°C. Greenhouses increased air temperatures by 3.22°C - 6.22°C in 2010 and 2.84°C - 3.37°C in 2011. Soil temperatures at 5 cm from the surface were only obtained below treeline in 2011, where mean temperature in the greenhouse was raised by 5.18°C. In 2012, mean soil temperatures inside vs. outside the greenhouses at the above and the below treeline sites differed less than 1°C while inside the greenhouse was 6.72°C higher than outside at the near treeline site.

Table 1: Comparisons of environment types at the study sites

Site		**Below treeline**	**Near treeline**	**Above treeline**
Elevation (m)		618	670	1169
Vegetation type		Spruce forest	Shrub tundra	Alpine tundra
Two dominant species (or growth forms)		Moss	Moss	*Dryas*
		Empetrum	Lichens	Graminoids
Summer 2011				
Air temperature mean (°C)	Ambient	n/a	10.36	9.27
	Inside greenhouse	n/a	13.73	12.11
Mean of daily maximum	Ambient	22.4*	18.60	15.03
	Inside greenhouse	31.39*	25.41	21.01
Mean of daily minimum	Ambient	6.07*	4.25	4.94
	Inside greenhouse	7.13*	5.63	6.20
Soil temperature mean (°C)	Ambient	3.42	n/a	n/a
	Inside greenhouse	8.60	n/a	n/a
2012 Soil temperature mean (°C)	Ambient	9.75	3.69	8.91
	Inside greenhouse	10.00	10.41	8.94
2010 Air temperature mean (°C)	Ambient	12.51	10.69	8.73
	Inside greenhouse	15.73	16.91	12.39
January 2011 Air temperature mean (°C)		−15.60	−13.13	−9.51
Light (μmol m-2 s-1) Sunny day	Ambient*	1656.50	n/a	1276.57
	Inside greenhouse*	1241.52	n/a	855.70
Light (μmol m-2 s-1) Cloudy day	Ambient*	376.86	512.17	500.58
	Inside greenhouse*	164.85	318.85	382.99
Air % Relative humidity (%)	Ambient	n/a	93.56	86.88
	Inside greenhouse	n/a	72.19	80.21
2012 Air Relative humidity (%)	Ambient	78.23	84.61	81.20
	Inside greenhouse	73.17	67.12	71.91
Soil % Volumetric water content	Ambient*	n/a	47.15	23.60

	Inside greenhouse*	n/a	35.00	19.50
Soil % Volumetric water content (Sampling)		53.42	52.49	20.36
Soil CN ratio		17.72	8.63	13.26
Soil texture		Sandy Loam	Loam	Sandy Clay Loam
Soil Bulk density (g/cm3)		0.182	0.795	0.559
Soil pH		6.0	5.0	6.5
Depth of thaw (cm)*		74.83	72.71	45.00

*Data 2010.

Three habitat types: above, near and below treeline sites. Data were obtained during a growing season (except January temperatures). Mean summer temperatures are recorded between 12-Jun-12-2011 and 29-Aug-2011 (the below treeline site was measured in 2010).

Okano and Bret-Harte

Okano and Bret-Harte *SpringerPlus* 2015 4:79, doi:10.1186/s40064-015-0833-x

Light response curve showed that seedlings' photosynthetic performance reached its maximum a little above 500 μmol m^{-2} s^{-1} light intensity, indicating that our measuring light intensity of 1500 μmol m^{-2} s^{-1} was beyond seedlings' light saturation at all sites.

Greenhouses blocked summer precipitation, which decreased air relative humidity (RH) and soil percent volumetric water content inside the greenhouse to 77.2% - 92.3% (similarly in 2012) and 74.2% - 82.6% of ambient levels, respectively (Table 1). The greenhouses also decreased light intensity on average to 68% (±2.78 SE) of ambient light levels depending on weather conditions; however, this reduction of light intensities was not statistically different (t-test, P=0.45). The inside greenhouse light intensity had reached over 1000 μmol m^{-2} s^{-1} (Table 1) at the below treeline site where the most shaded condition was expected among the sites. Light intensity in the greenhouses was always above the levels that caused photosynthetic light saturation (a little above 500 μmol m^{-2} s^{-1}; on sunny days. This suggested that seedlings inside the greenhouses at all sites were not shade-adapted. On cloudy days, light levels in greenhouses were below saturation (Table 1). However, all photosynthetic measurements were made at saturation (1500 μmol m^{-2} s^{-1}). Although caution is needed in interpreting data, we believe that the use of the greenhouses did not affect our photosynthesis results.

Seedling Size

For several variables, results of three-way ANOVA indicated that there were significant interactions between habitat type and temperature, and between temperature and removal treatments (Table 2). High temperatures influenced the growth response (relative growth rate, RGR) of both height and number of needles of *P. glauca*, but not in the same way at each habitat. Above treeline, both height and number of needles were increased significantly by elevated temperatures. Near treeline, height was increased with marginal significance by high temperatures, but the number of needles was decreased, although the differences between high and ambient temperature treatments were not significant. Below treeline, height and number of needles were decreased when temperatures were high, but not significantly.

Table 2: Results of statistics (general linear model, GLM)

Factor	Habitat type (H)		Temperature (T)		Removal (R)		H*T		H*R		**T*R**		**H*T*R**	
	F	**P**	**F**	**P**	**F**	**P**	**F**	**P**	**F**	**P**	**F**	**P**	**F**	**P**
Size measurements	Ndf: 2	Ddf:144	Ndf: 1	Ddf:144	Ndf: 1	Ddf:144	Ndf: 2	Ddf:144	Ndf: 2	Ddf:144	Ndf: 1	Ddf:144	Ndf: 2	Ddf:144
(1) RGR Height†	9.51	0.0001	11.26	0.001	0.23	0.633	15.43	<0.0001	1.96	0.145	5.05	0.026	1.43	0.244
(2) RGR # of needles†	28.14	<0.0001	1.79	0.183	9.22	0.003	3.70	0.027	1.47	0.232	0.50	0.480	0.71	0.493
(3) RGR Needle length‡	7.10	0.0011	0.19	0.664	0.61	0.438	1.72	0.182	0.18	0.837	0.06	0.807	0.97	0.380
Photosynthesis	Ndf: 2	Ddf:143	Ndf: 1	Ddf:143	Ndf: 1	Ddf:143	Ndf: 2	Ddf:143	Ndf: 2	Ddf:143	Ndf: 1	Ddf:143	Ndf: 2	Ddf:143
(4) Photosynthesis*	32.43	<0.0001	0.01	0.9312	0.44	0.510	52.92	<0.0001	1.24	0.2916	4.22	0.042	0.65	0.523
% N and ^{13}C	Ndf: 2	Ddf:47	Ndf: 1	Ddf:47	Ndf: 1	Ddf:47	Ndf: 2	Ddf:47	Ndf: 2	Ddf:47	Ndf: 1	Ddf:47	Ndf: 2	Ddf:47
(5) %N**	5.80	<0.0001	0.12	0.735	7.19	0.010	1.91	0.160	1.55	0.223	0.05	0.826	1.13	0.333
(6)^{13}C	8.01	0.001	10.27	0.002	0.39	0.536	4.80	0.013	0.44	0.646	0.85	0.361	0.17	0.200
ANCOVA	Ndf: 2	Ddf:46	Ndf: 1	Ddf:46	Ndf: 1	Ddf:46	Ndf: 2	Ddf:46	Ndf: 2	Ddf:46	Ndf: 1	Ddf:46	Ndf: 2	Ddf:46
(7) %N**	5.95	0.0051	0.38	0.543	0.13	0.719	37.58	<0.0001	1.15	0.324	0.44	0.512	0.39	0.680
(8)^{13}C	16.57	<0.0001	0.60	0.441	0.50	0.481	36.87	<0.0001	0.91	0.409	0.57	0.456	0.19	0.832

†rank, ‡yt=y/s, *log(y+1), and **log (y) transformation to achieve homogeneity of variance. (Bold: $P<0.05$).

Three-way ANOVA for relative growth rates (RGR) in (1) height, (2) number of needles, (3) length of the longest needle for size measurements; (4) photosynthetic rate, (5)%N and (6) $\delta^{13}C$. ANCOVA for photosynthetic rate with (7)%N and (8) $\delta^{13}C$. The interactions between the treatments were shown; e.g., H*T (habitat type and temperature). Significant differences were bold. Data were collected from 156 seedlings (one failed for photosynthesis, n=155). ANCOVA was run using a subset of seedlings (n=59). Transformation methods were suggested by Dr. J. McIntyre of Department of Mathematics and Statistics, UAF.

Okano and Bret-Harte

Okano and Bret-Harte *SpringerPlus* 2015 4:79 doi:10.1186/s40064-015-0833-x

Seedlings in higher temperatures increased heights significantly above treeline ($F_{(1, 51)}=40.17$, $P<0.0001$, one-way ANOVA) and marginally significantly near treeline ($F_{(1, 54)}=3.06$, $P=0.086$, one-way ANOVA). Although the difference below treeline was not significant ($F_{(1, 45)}=2.00$, $P=0.16$, one-way ANOVA), warmed seedlings were 30% shorter without neighbors and 45% shorter when neighbors were present relative to seedlings in the ambient temperature. There was a significant interaction between temperature and removal for RGR in height because there were increased competition effects overall when the temperature was higher (Table 2, Figure 1a).

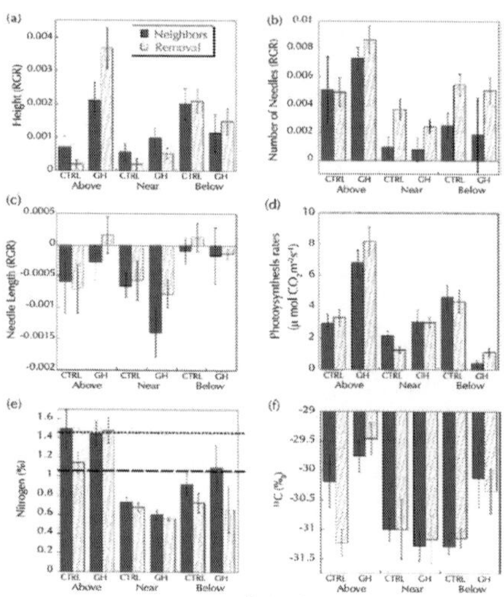

Figure 1: Results of growth response of*Picea glauca*. Relative growth rates (RGR) of (a) height, (b) number of needles, and (c) length of the longest nee-

dle (n=156). (d) Gross photosynthesis rate (n=155). (e) Percentage of N content (n=59). The upper dotted line is an adequate nitrogen level (1.45%). The lower dotted line is a severe deficiency level (1.05%). (f) $\delta^{13}C$ content (n=59). CTRL=control temperatures. GH=elevated temperatures. Above=above treeline site, Near=near treeline site, Below=below treeline site. Error bars show one standard error.

Neighbor removal increased number of needles at all sites but one (control above treeline, insignificant difference) and there was no significant interaction with temperature (Table 2, Figure 1b). The number of needles increased with higher temperatures above treeline ($F_{(1, 51)}$=6.97, P=0.01, one-way ANOVA), but there were no significant effects at the other two sites. Length of the longest needle decreased over the summer due to leaf turnover. This decrease in leaf length was significant only for those at the near treeline site (P=0.001, Table 2, Figure 1c).

Overall, warming affected the RGR of seedlings positively above treeline, but had slightly positive to weakly negative effects at the lower two habitat types.

Photosynthetic Response

There was a significant interaction between temperature and habitat type (P<0.0001, Table 2), because elevated temperatures significantly increased photosynthetic capacity (photosynthetic rate at 1500 μmol m^{-2} s^{-1}) above ($F_{(1, 51)}$=33.93, P<0.0001, one-way ANOVA) and near ($F_{(1, 54)}$=10.80, P=0.0018, one-way ANOVA) treeline, but significantly decreased photosynthetic capacity below treeline ($F_{(1, 45)}$=49.82, P<0.0001, one-way ANOVA). Although post hoc Tukey HSD tests were not significant, the interaction between temperature and removal effects were significant ($F_{(1, 143)}$=4.22, P=0.042), because competition at the near treeline site became marginally more pronounced under elevated temperatures than under ambient temperatures ($F_{(1, 54)}$=3.88, P=0.054, two-way ANOVA, Figure 1d).

Foliar N and C, and their Relationships with Photosynthesis

Foliar N and C were indicators of seedlings' nitrogen availability and drought stress, respectively. Nitrogen concentration of needles was significantly higher above treeline where photosynthetic rates were also high (Table 2, Figure 1e). There, the foliar %N was at a level previously suggested to be adequate for growth (1.45%, Carter [1992]). All of the seedlings near treeline and most of the ones below treeline had nitrogen levels indicative of severe deficiency 1.05%, (Carter [1992]). Except in one treatment, seedlings with neighbors had significantly higher nitrogen concentration in needles than did ones without neighbors at the same site (Table 2, Figure 1e). $\delta^{15}N$ values of *P. glauca* have been observed to range between −1‰ and −11‰ due to variations in nitrogen availability to plants and their relationships with mycorrhizal fungi that supply organic nitrogen; hence, %N and $\delta^{15}N$ are often correlated (Hobbie et al. [2009]). In our study, all but one (−0.80‰) of the 59 subsampled seedlings had a positive value of $\delta^{15}N$ (+2.91‰±0.19SE, data not shown), and %N did not correlate with $\delta^{15}N$

With respect to C, $\delta^{13}C$ levels in needles of seedlings growing above treeline were less negative (more enriched) than the levels in needles from other habitat types (Figure 1f). Overall, $\delta^{13}C$ levels under ambient temperatures were more negative (more depleted) than the ones under elevated temperatures. This could indicate a small amount of water stress in the greenhouse. However, this difference was only significant above treeline. All seedlings had $\delta^{13}C$ values of −29.42‰ ~ −32.81‰, which is within the range or even slightly more negative than previously published data for *Picea* (−26‰ to −31‰; Grossnickle [2000], or −29‰ for *P. glauca*; Livingston et al. [1999]). This indicates that no seedlings suffered from severe water stress, that they could open their stomata in order to obtain CO_2, and could discriminate against heavy ^{13}C during needle production.

In order to assess whether foliar %N or $\delta^{13}C$ explained the differences in photosynthetic performance, we ran ANCOVAs, using photosynthesis data as a dependent variable and either %N or $\delta^{13}C$ as covariates. In both cases, the covariates eliminated the interaction effects between temperature and removal (Table 2), but not the habitat by temperature interactions and the main effects of habitat type. Variation in %N or $\delta^{13}C$ partially explained the significant effect of habitat on photosynthetic rates, but did not make it insignificant in the ANCOVA, suggesting that additional factors that differ among habitats were important.

DISCUSSION

Elevated temperature increased growth and photosynthetic rates most strongly above treeline and slightly near treeline, but decreased them below treeline, which indicates that warmer temperatures have different effects depending on habitat. Climate warming can cause drought stress on plants, and such unfavorable abiotic conditions commonly trigger facilitative effects in a plant community. However, our $\delta^{13}C$ results suggest that temperature-induced drought stress was not an important factor in our experiment, and we saw no evidence for facilitation, suggesting that climate warming will have a positive effect on the growth of seedlings where it is currently too cold for *P. glauca*. Warmer temperatures increased competition between spruce seedlings and neighboring plants at all sites.

Above treeline, where seedlings did not suffer from drought stress, photosynthetic rates were increased by elevated temperatures. The seedlings above treeline, both inside and outside the greenhouse, had foliar nitrogen levels close to what is considered an adequate level (1.45%, Carter [1992]), yet the seedlings outside at ambient temperatures did not photosynthesize as much as the seedlings at high temperatures, indicating that higher temperatures increased photosynthesis rates when N was not limiting. Higher temperatures could increase either activity or production of Ribulose 1,5-bisphosphate carboxylase/oxygenase (Rubisco) and other Calvin cycle enzymes, which would amplify carboxylation capacity. Plants tend to produce more Rubisco than needed, particularly when nitrogen is abundant (Stitt and Schulze [1994]). The high foliar nitrogen concentration suggests abundant Rubisco. Therefore, it appears that an increased amount of Rubisco allows the seedlings under high temperatures to fix more carbon and allocate enough photosynthates for both height and needle growth.

Seedlings near treeline also increased their photosynthetic rates significantly under warming relative to ambient temperatures at the same habitat type, although photosynthetic rates near treeline were the lowest of all sites. Unlike seedlings above treeline, warmed seedlings allocated more carbon to height growth than to needle production. In the ambient temperature, low photosynthetic rates at the near treeline site may have occurred because low soil temperatures induce stomatal closure and inhibit photosynthesis (Goldstein et al. [1985]) and can suppress the height growth of *P. glauca* (Zhang and Dang [2007]). Near treeline was not invaded by spruce until recently (Hamm [2007]). Mineral soil just below the organic soil (a depth of 15–20 cm) was frozen at the end of May and soil quality differs from other sites (Table 1), which might affect soil turnover and nutrient availability, and subsequently establishment of spruce. The increased soil temperature caused by the greenhouse (Table 1) could increase the photosynthetic rates and promote seedlings' growth. However, because these seedlings were nutrient-limited, there may have been a trade-off between above and belowground growth, which may have resulted in less carbon available for aboveground growth and needle production than at the above treeline site. Alternatively, allocation to height over needle production may have resulted from competition for light. This allocation pattern has been seen previously in *Picea asperata*; elevated temperatures significantly increased height and root mass ratio, but decreased leaf mass ratio for *P. asperata* seedlings (Yin et al. [2008]). Perhaps a lack of needle production prevented the near treeline seedlings from a rapid increase in photosynthetic performance, and hence, from a rapid increase in growth. Still, recent warming temperatures have a potential to promote *P. glauca* seedlings by increasing photosynthesis and height growth near treeline.

Trade-offs are important for the survival of plants in infertile environments, which are common in the Arctic and subarctic. Instead of partitioning all photosynthates to tissues that support growth, slow growing species invest in storage and traits that conserve resources (Grime et al. [1997], Grime [2001], Díaz et al.[2004]). Alternatively, plants may trade-off resource allocation between roots and shoots (Tilman [1988]). Conifers, including spruce, tend to allocate photosynthates toward belowground biomass when either soil nutrients or water are scarce (Waring [1987], Grossnickle [2000]). In some places, positive relationships between aboveground net primary productivity and belowground net primary productivity have been seen, but in boreal forests, there is often a negative relationship (Litton and Giardina

[2008]). In boreal forests, there is also a negative relationship between the mean annual temperature and total belowground carbon flux, presumably due to changes in resource allocations with increasing nutrient availability caused by rising temperatures (Litton and Giardina [2008]). While our experiment focused on aboveground allocation only, as temperature increased, the seedlings showed differences in allocation by habitat type, indicating differences in their growth strategies. Seedlings above treeline increased total photosynthates, thereby increasing in height and the number of needles, perhaps also increasing in belowground biomass. Seedlings near treeline apparently allocated photosynthates mainly toward height and possibly some toward belowground structures, presumably due to lack of nutrient availability as indicated by their low foliar N (Waring [1987], Grossnickle [2000]).

In contrast to the other two habitats, warming affected seedlings negatively below treeline; there, elevated temperature decreased photosynthetic capacity (Figure 1d). The below treeline site is a mature spruce forest whose current environment is favorable for mature *P. glauca* trees. However, if the current temperature regime is not the same as when these trees were established, and it is near the upper limit of the species' tolerance now, a slight increase in temperature could make the habitat unfavorable to the species. During the 2010 growing season, both the average temperature and the average maximum temperature in a greenhouse below treeline were much higher, by 3°C and 9°C, respectively, than the ambient temperatures (Table 1). The optimal temperatures for photosynthesis of *P. glauca* seedlings have been reported as 15°C or 25°C (Man and Lieffers [1997]). Mature trees of *P. glauca* in the Alaska Range decreased their growth when the average July temperature in Fairbanks reached 16°C or higher, which on average corresponds to 11.5°C at the Headquarter of DNPP (Wilmking et al. [2004]). The average temperature in the greenhouse below treeline (15.7°C) was greater than this threshold (Wilmking et al.[2004]), and the average maximum temperature (31.1°C) is also too high for the optimal temperature in photosynthetic performance of seedlings (Man and Lieffers [1997]). Our results suggested that the current habitat of mature spruce forests will likely become less favorable to both mature and juvenile *P. glauca*species as climate warms in the future.

Although negative effects of warming on *P. glauca* have been reported by previous studies (e.g. Barber et al. [2000], Wilmking et al. [2004]), considering the species' distribution, our result that the environment of below treelines would become unfavorable to *P. glauca* due to elevated temperatures was surprising. *P. glauca* occurs from Alaska to South Dakota, Wyoming and northern New England, and the southern limits fall into the 18°C July temperature isotherm (Nienstaedt and Zasada [1990]). Alaska is near the northern edge of its distribution and why a species that can reside in more southern areas has suffered from summer warming in Alaska is puzzling. One possible explanation is due to ecotypic genetic differentiation between Alaskan and non-Alaskan *P. glauca*. Alaskan *P. glauca* had a unique haplotype of chloroplast DNA that distinguished it from *P. glauca* in northeastern North America (Anderson et al. [2006], de Lafontaine et al.[2010]). *P. glauca* in Alaska was assumed to have been able to survive the last glacial maximum in the Beringia refugium, not only due to the local microclimate, but also due to being adapted to severely cold climate (Anderson et al. [2006]). Perhaps the seedlings in our experiment originating in DNPP are genetically better adapted to cold temperature more than *P. glauca* from different origins, and are more sensitive to warmer temperatures.

Our growth and photosynthesis results suggest that interactions between *P. glauca* and neighboring plants were mostly competitive, and that competition was aggravated under elevated temperatures. Seeing strong competitive effects above treeline contrasts with previous alpine studies that concluded facilitation was more important than competition between plants under a harsh abiotic environment (Callaway [1998], Cavieres et al. [2002]). Limiting factors for the growth of plants at high altitudes include non-resource factors such as temperatures, winds and disturbance of soil rather than resources such as soil nutrients; the opposite is true at lower altitudinal sites (Callaway et al. [2002]). At the two higher sites (above and near treeline), rising temperatures eliminated one of the non-resource limiting factors, so soil resource may have become a stronger limiting factor. On the other hand, the seedlings below the treeline site, the current habitat of a mature forest, recorded negative effects of elevated temperatures on the seedlings, but again no facilitation effects were seen. Where facilitation develops in harsh environments, there is likely a trade-off between the damaging effect of winter winds, which is lessened by being close to neighbors, and reduced nutrient availability due to uptake by neighbors. Because no physical damage by winds occurred in the mature forest, there was no obvious advantage there for seedlings to have neighboring plants. Seedlings below treeline were mostly under severe nitrogen deficiency, so belowground competition was presumably the limiting factor for the below treeline seedlings, as Callaway et al. ([2002]) suggested.

Our experiment suggested that competition and nutrient availability may limit growth of the *P. glauca*seedlings under warming even though warming has a positive effect. This result is consistent with previous*P. glauca* experiments. Warming increased the growth of transplanted *P. glauca* seedlings for more than three years in arctic tundra sites and belowground competition limited growth there, presumably due to nutrient competition (Hobbie and Chapin [1998]). Effects of competition with *Populus tremuloides* (aspen) on the growth of *P. glauca* increased more than proportionally when temperatures rose (Cortini et al.[2012]). *P. glauca* seedlings in boreal forests did not obtain any benefits from neighboring shrubs and herbaceous plants (Cortini and Comeau [2008]). Sullivan et al. ([in press]) found evidence that the growth of *P. glauca* was increased by increased nutrient availability due to warmer soil temperatures, rather than air temperature. These results are similar to what we found and suggest that competition, especially belowground competition between spruce seedlings and neighboring plants, could limit the growth of *P. glauca* seedlings when temperatures increase in the future. Since our study focused on aboveground allocation, further studies on seedlings' resource allocation including belowground allocation are needed to support this interpretation. If the negative effect of competition becomes more pronounced than the positive effect of warming, expansion of spruce habitat in the future may not occur so rapidly as treeline expansions have proceeded for the past 50 years in DNPP (Stueve et al. [2011], Hamm [2007]).

Our study centered on seedlings of *P. glauca* at treeline ecotones, while many other studies have focused on mature trees of this species. Because successful recruitment and establishment of seedlings are essential for expansion of treeline, studying the response of seedlings to changing climate is necessary. Most importantly, seedlings are generally more vulnerable than mature trees and likely more sensitive to environmental changes such as warming temperatures. Successful recruitment years may be followed by decades with no recruitment at treeline as in Russia (Gervais and MacDonald [2000]). Similar observations of pulses of recruitment have been made in Sweden (Kullman [1991], Dalen and Hofgaard [2005]). This indicates that seedlings are sensitive to interannual variability in climate, which affects their recruitment, establishment, or mortality. However, trees might need only a few summers of extraordinarily good weather for recruitment to keep up the species' population (Körner [2012]). Our study focused on establishment and growth of surviving seedlings, and found that habitats above and near treeline have a potential to promote the growth of spruce seedlings as climate warms. Further studies of seed viability and germination, growth of older seedlings and saplings, and causes of mortality of seedlings in the field before and after establishment would also improve our understanding of seedlings' growth capacity at treeline ecotones in the warmer future.

CONCLUSIONS

Our results suggest that increasing temperatures would favor the growth of *P. glauca* seedlings at those places where it is too cold now. On the other hand, it is questionable whether Alaskan *P. glauca* would be able to continue to grow in its current habitat below treeline as climate warms. Current mature forests could become unfavorably warm for seedling growth of *P. glauca*. Warming may also promote increased competition, likely belowground, which could slow rates of habitat expansion of *P. glauca* above treeline. *P. glauca* is likely to remain the dominant treeline species, and treeline expansion will keep occurring in this subarctic mountain region as the climate warms, but may not be as fast as we have seen during the past 50 years.

MATERIALS AND METHODS

Study Area Description

The study area was located in Camp Denali and North Face Lodge, one of the private inholdings within DNPP near the southwestern end of the Kantishna Hills approximately 46.7 km north of the Alaska Range (Hamm [2007]). The closest weather station is approximately

3 km away at Wonder Lake (63° 29′25″N, 150° 52′17″W, elevation 646 m, which is between our near treeline and below treeline sites. Western Regional Climate Center. http://www.wrcc.dri.edu) indicated that mean temperature, mean humidity, and total precipitation during the experimental period in 2011 (June 12 - August 29) were 10.75°C, 75.78% and 205.46 mm, respectively. The Wonder Lake temperature was similar to that measured in our near treeline site (Table 1). Three experimental sites were established, at different habitat types relative to the current treeline: a site above, near, and below (Table 1). At each site, two plots (on relatively flat ground no more than 5 m apart) were randomly chosen within vegetation representative of that where *P. glauca* occurred in the area (if it occurred).

The highest, above treeline site (63°32′17″N, 150°53′24″W, elevation 1169 m) was on an exposed hilltop with no aspect and no shade (Figure 2a). The area was covered with subarctic and alpine tundra species, but rocks and soil were also exposed. Vegetation of the study area included evergreen (*Cassiope tetragona,Loiseleuria procumbens, Rhododendron lapponicum, Dryas octopetala*) and deciduous (*Salix* spp.) tundra shrub species, forbs (*Primula cuneifolia, Anemone narcissiflora, Oxytropis nigrescens*), graminoids (*Carex*spp), and moss and lichen species.

Figure 2: Experimental sites. (a) Above treeline. (b) Near treeline. (c) a sapling found at the near treeline site. (d) Below treeline.

The near treeline site (63°31′09″N, 150°53′52″W, elevation 670 m) was on a south-facing slope in open shrub tundra with occasional, scattered *P. glauca* saplings (Figure 2b, c). It was lower than altitudinal treelines elsewhere in the Alaska Range (up to 1,550 m, Sveinbjörnsson [2000]), possibly because of poorly drained soils underlain by permafrost. Recent climate change, however, has promoted white spruce establishment and growth in this area for the past 50 years (Hamm, [2007]). Treeline is defined not only by elevation (Sveinbjörnsson [2000]); considering heterogeneity of treeline including topology, interaction with neighboring vegetation and local climate that affect treeline, we followed the broad definition of treeline as Stueve et al. ([2011]) employed, which made us define this site as "near treeline tundra." There were very few juvenile *P. glauca* saplings in this area (Figure 2b,c). Those that could be measured had an average age between 18 and 19 years (n=3), based on counting annual extension growth. Most of the area was open and vegetation was composed of evergreen (*Vaccinium vitis-idaea, Rhododendron subarcticum* ([USDA, ARS, National Genetic Resources Program]; formerly known as *Ledum palustre* ssp. *decumbens*), *Empetrum nigrum*) and deciduous (*Betula nana, Vaccinium uliginosum, Rubus chamaemorus*) shrub species, graminoids (*Carex* spp), lichens and mosses. The moss layer, mostly composed of*Sphagnum* spp., was between 5 cm and 10 cm deep, and was heavily mixed with lichens.

The below treeline site (63°32′10″N, 150°54′24″W, elevation 618 m) was located on a south-facing slope in a mature white spruce forest (Figure 2d). We cored four trees (randomly selected) near our site. The tree cores gave an average age of 127.4 years after correcting for time to grow to core height (Lloyd and Fastie[2003], Hamm [2007]), although this may not represent age of the entire stand. The study site was not a closed canopy, but was sunny with occasional shading by mature trees (*P. glauca, Populus balsamifera*). Vegetation was composed of evergreen (*Juniperus communis, V. vitis-idaea, Rhododendron groenlandicum*(formerly known as *Ledum palustre ssp. groenlandicum*), *Empetrum nigrum*) and deciduous (*V. uliginosum*) shrub species, and a thick (more than 20 cm deep) layer of mosses.

Seedling Preparation

Seedlings of *P. glauca* were grown in a greenhouse at University of Alaska Fairbanks (UAF). Seeds in cones were harvested in DNPP from three elevations, 640 m, 675 m, and 725 m between August 29 and September 1, 2008 by Dr. C. Roland of the National Park Service. The cones were stored at a temperature of −26°C. In December 2009, spruce seeds were placed onto moist filter papers in petri dishes and stratified at 4°C for more than sixty days. In March 2010, seeds were moved into a growth chamber at a temperature of 21°C and given 14 hours of light each day to trigger germination (germination rate 38.0%). Each germinated seed was planted in soil (greenhouse standard media) in a pot measuring 4 cm diameter and a depth of 20 cm. Seedlings were watered with distilled water every other day for 10 days, then 10–30 - 20 (Nitrogen 10%, Phosphorus 30%, and Potassium 20%) fertilizer was applied at 50 mg/L N for the first 10 days, followed by 100 ppm N fertilizer for a week and 200 ppm N for about 45 days. Of 320 planted seeds, 53.4% survived (171 seedlings). Seedlings were assigned to treatments (as described below; Table 3) so that the seedlings with the same origin/similar size were evenly spread throughout treatments. At $\alpha=0.05$ significance level, mean height, number of needles, and the longest needle in each treatment group when planted were not significantly different among treatments ($F_{(11,159)}=1.702$, $P=0.077$, $F_{(11,159)}=1.210$, $P=0.284$, $F_{(11,159)}=1.690$, $P=0.079$, respectively).

Table 3: Treatments, seed origin, original size and number of survivals of seedlings in September 2011

Habitat	Temperature	Neighbors	Site#	n	Origin 1	Origin 8	Origin 9	Height (cm)	#Needles	Length (cm)	Dead	Survived n
Above	Control	Control	a	15	3	0	12	4.56 (1.60)	136.6 (53.1)	1.47 (0.20)	4	11
treeline	Control	Removal	b	14	3	0	11	4.49 (1.52	150.0 (61.0)	1.54 (0.19)	0	14
(1169 m)	High	Control	c	14	3	0	11	4.00 (1.43)	118.4(53.5)	1.49 (0.19)	0	14
	High	Removal	d	14	3	0	11	4.17 (1.85)	113.1 (46.9)	1.46 (0.12)	0	14
	Above site total/average			57				4.31 (1.60)	129.5 (53.6)	1.49 (0.18)	4	53
Near	Control	Control	e	15	3	0	12	4.02 (1.07)	162.2 (64.5)	1.55 0.23)	0	15
treeline	Control	Removal	f	14	3	0	11	4.86 (1.53)	173.9 (82.1)	1.61 (0.21)	0	14
(680 m)	High	Control	g	14	3	0	11	4.77 (1.13)	142.2 (43.2)	1.59 (0.21)	1	13
	High	Removal	h	14	3	1	10	5.69 (0.84)	146.6 (42.2)	1.59 (0.17)	0	14
	Near site total/average			57				4.84 (1.14)	156.23 (58.0)	1.59 (0.21)	1	56
Below	Control	Control	k	15	3	1	11	4.13 (1.30)	139.5 (54.3)	1.39 (0.16)	2	13
treeline	Control	Removal	l	14	3	1	10	4.36 (1.27)	156.9 (70.6)	1.47 (0.21)	0	14
(617 m)	High	Control	i	14	3	1	10	4.94 (1.88)	155.3 (78.8)	1.49 (0.19)	8	6
	High	Removal	j	14	3	1	10	4.96 (1.44)	162.4 (67.2)	1.55 (0.20)	0	14

	Below site total/average	57		4.60 (1.47)	153.5 (67.7)	1.48 (0.19)	10	47
	Total/Average	171		4.58 (1.41)	146.4 (59.8)	1.52 (0.19)	15	156

A total of 12 treatments were tested on 171 white spruce seedlings. After two summers, 156 (or 91%) of seedlings survived. In the "High" treatment, temperature was increased by small greenhouses. Above ground neighboring plants were removed in "Removal" plots. "Origin" is where the cones of the seeds were collected. "Origin 1" = 675 m, "Origin 8" = 725 m, "Origin 9" = 640 m. Mean "Height," number of needles "#Needles," and length of the longest needles "Length" (+SD) were measured at the time of planting. "Dead" and "Survived n" were recorded in September 2011.

Okano and Bret-Harte

Treatments

The spruce seedlings were planted at the study area between June 2 and 11, 2010, under 12 treatments, consisting of 3 habitat types, 2 temperature levels, and 2 competition levels (Table 3). Three habitat types were: 1) above treeline on an exposed ridge with alpine tundra vegetation; 2) near treeline in shrub tundra; and 3) below treeline in a mature *P. glauca* forest (control); as described above.

Two plots per site were established whose temperature levels were either 1) elevated temperatures to simulate climate change or 2) ambient temperatures (control). In order to passively increase air temperatures, small greenhouses were used in each habitat type during the summer seasons. Greenhouses were put into place in an early June and disassembled in late August of each year. An A-frame greenhouse (Figure 3a) constructed of 140 cm × 120 cm rectangular wooden frames with a height of 85 cm was covered with transparent 0.15 mm (0.6 mil) plastic sheeting. Top parts of the front and back walls were loosely opened for air ventilation. Through two growing seasons, air temperatures and relative humidity (RH, only in 2011) were recorded using temperature loggers (iButtons, DS1921, DS1923; Maxim Integrated, San Jose, CA). One logger was placed at the center of a greenhouse or an outside site for taking air measurements. Another logger for soil temperatures was buried just below the air logger.

Figure 3: Treatments. (a) Greenhouse (elevated temperatures) and outside (ambient temperatures = control) sites above the treeline sites. (b) Removal (presence of competition) and non-removal (absence of competition = control) sites.

In previous studies, no effect of greenhouses on soil moisture was reported (Bret-Harte et al. [2002], Rixen and Mulder [2009]) despite a lack of precipitation inside. The effect of winds could have been minimized by a greenhouse during three months of the summer growing season, but not during the majority of the year (9 months) after the greenhouse was disassembled in September, including winter when physical damage of seedlings due to wind was mainly expected.

Two levels of competition (Figure 3b) were 1) the absence of competition - removal of neighboring plants and 2) the presence of competition - neighboring plants present (control). Each seedling was planted at the center of a 15–18 cm diameter plot (Choler et al. [2001]), which was large enough that seedlings whose neighbors were removed would not be shaded by surrounding plants, with 2–5 cm of buffer zones between seedlings. One seedling was planted with naturally occurring neighbors in the plot, while the next seedling was planted in a plot where aboveground biomass of neighboring species was removed by clipping. Belowground biomass was not removed, but the border soil between the plot and outside vegetation was cut to a maximum depth of 25 cm with a knife to minimize belowground competition. This condition was maintained by clipping the regrowth of neighboring plants in June and September each year. If competition were occurring, seedlings would grow better without neighbors. If facilitation were occurring, seedlings would grow better with neighbors.

Our experiment was conducted at a remote location with limited access, which limited what we could do. The most important limitation was that we could only have one greenhouse and one outside plot per site, in which 14–15 replicated seedlings were planted. This design limits our ability to extrapolate across the landscape, even though we established our plots on representative locations for each habitat type. Soil moisture was not statistically different between greenhouse and control plots in the same site where we could measure it (Table 1; t test, $P=0.45$ above treeline, $P=0.49$ near treeline). Given the constraints of doing research in a private inholding in the National Park, and the need to conduct measurements over a short period of time to be biologically reasonable, we felt this design was the best compromise.

Data Collection and Analyses

After two summers and one winter at the end of the growing season of 2011, data were collected from 156 surviving seedlings (survival was 91%). Analyses included non-destructive size measurements, maximum gross photosynthetic rates, and concentrations of foliar nitrogen (N), carbon (C), and the natural abundances of ^{15}N, and ^{13}C (expressed as $\delta^{15}N$ and $\delta^{13}C$, respectively) from a subset of samples.

Size data was obtained between August 27 and 31, consisting of height, number of needles and the longest leaf length of each seedling. Based on size measurements in May 2011, relative growth rates (RGR) that were used for statistical analyses were calculated for a comparison as follows:

$RGR=(\ln Size_{final}-\ln Size_{initial})/(day_{final}-day_{initial})$....Durations=90~93 days

Maximum photosynthetic capacity of seedlings was measured between 11:50 am and 5:25 pm from August 28 to September 3. For the seedlings in a greenhouse, measurements were taken on the day when the greenhouse sheeting was removed. Net photosynthesis and dark respiration rates were measured from all but one seedling (one failed, n=155) using a LiCor 6400XT portable photosynthesis system (LiCor, Lincoln, NE, USA) under the same CO_2 concentration of 400 ppm. Maximum net photosynthesis rates were recorded every 2 seconds for one minute (30 records) after placing a sample chamber over a seedling 2 cm from its top and exposing the seedling to a light intensity of 1500 μmol m^{-2} s^{-1} (full sunlight level) for two minutes. This shoot included a mixture of new and old needles. Just after photosynthesis was recorded, the light was turned off (0 μmol m^{-2} s^{-1} light intensity). After one minute of seedling acclimation to darkness, 30 dark respiration records were measured with the same time interval. Each set of 30 records was averaged and used to calculate an average gross photosynthesis rate per seedling as follows:

Gross photosynthesis=Net photosynthesis(a positive value)-Dark respiration(a negative value)

We assume that measurements in the dark represent respiration in the light, although we acknowledge that gross photosynthesis defined in this way possibly overestimates respiration because of the Kok effect, the inhibition of respiration in the light (Kok [1948], Heskel et al. [2013]). However, very little difference between day and night time respiration of black spruce seedlings was seen previously (Way and Sage[2008]).

We wanted to know whether the reduction of the light intensity by the greenhouse sheeting was substantial or negligible for seedlings' photosynthetic performance. In order to estimate the maximum capacity of photosynthetic performance, a light response curve was created in March 2011, using 9, two-month old seedlings raised in a greenhouse from the same set of cones originating from Denali, with the same protocol stated previously. In addition, in the field, light levels inside and outside the greenhouses were measured using a light sensor on the LiCor 6400XT photosynthesis system. The sensor was placed facing directly to the sun light and data were recorded every 2 seconds for one minute. Data were obtained both sunny and overcasting days, obtaining 2–4 data per either inside or outside a greenhouse, then averaged.

Leaf area of each seedling was estimated by establishing a relationship between the total length of needles held in the sample chamber and their area. The needles used by photosynthesis measurements could not be harvested because this would damage the seedlings. Instead, needles of a-month-old seedlings that were from the same sets of cones but not planted in the field, were harvested prior to the field experiment. Leaf length and area of all harvested needles were measured and a linear regression between leaf area and leaf length was constructed:

Leaf area(cm^2)=0.0781*L*N

Where *L* is the average length of the needles and *N* is number of needles used. In the field, the number of needles 2 cm from the top of each seedling was counted, and the lengths of four representative needles were measured. Area was estimated from the length of those needles.

In order to investigate the relationship of foliar N and C with photosynthesis rates, 10–15 needles were collected from a subset of seedlings (n=5 per treatment, 59 total due to failure of a photosynthetic rate) after each size measurements. Stable isotope fractionation in $\delta^{15}N$ and $\delta^{13}C$ was focused to explain plants' dependency in nitrogen on mycorrhizal association ($\delta^{15}N$) and plants' drought stress level ($\delta^{13}C$). Samples were dried at 60°C and ground. Concentration of N, C, $\delta^{15}N$ and $\delta^{13}C$ were determined by continuous flow isotope ratio mass spectrometry using a Costech ECS4010 Elemental Analyzer (Costech Analytical Technologies Inc.,Valencia, CA) interfaced to a Finnigan Delta plus XP isotope ratio mass spectrometer via Conflo III interface (Thermo Fisher Scientific Inc., Waltham, MA) at the Alaska Stable Isotope Facility at UAF.

Other environmental data were obtained either in the fall of 2010 or 2011. Three soil samples per site (<5 m from the plots) were collected by cutting soil with a knife and measured the volume. Each sample was dried at 105°C over 3 days and weighed to calculate a bulk density and water content. Another sample per site was collected for CN, pH, and texture tests. The percentage of C, N contents of soil was analyzed at the Forest Soils Laboratory at the UAF School of Natural Resources and Extension. Soil pH was tested using a pH test indicator (General Hydroponics-USA, Sebastopol, CA, USA). Texture was defined by Bouyoucos Hydrometer Method. In addition to sampling, soil volumetric water content was measured using a Hydrosense soil water measurement system (Cambell Scientific Inc., North Logan, UT, USA).

Data on size measurements, gross photosynthesis rates, and foliar CN data were analyzed using three-way ANOVA (general linear model (GLM) with habitat, temperature and competition as main effects and all possible interactions). Post-hoc tests (Tukey's HSD) were used to interpret significant effects. A subset of the photosynthesis data was also analyzed by ANCOVA. Either N concentration or $\delta^{13}C$ was used as a separate covariate in order to assess whether foliar %N or fractionation of $\delta^{13}C$ explained the differences in photosynthetic performance. Data were tested for homogeneity of variance before the analysis using four tests (O'Brien, Brown-Forsythe, Bartlett and Levene); if necessary, data were transformed using ranking, weighted standardization, or log transformation as indicated in Table 2. When interaction terms were significant, data were further analyzed using one-way ANOVA.

AUTHORS' CONTRIBUTIONS

KO originally formulated the idea. KO and MSB designed the experiments. KO conducted fieldwork. KO and MSB analyzed the data. KO wrote the manuscript, with assistance from MSB. Both authors read and approved the final manuscript.

ACKNOWLEDGMENTS

We appreciate the help of Dr. G. Laursen, Dr. M. Wright, Dr. C. Roland, Dr. J. Alden and others at the University of Alaska Fairbanks regarding scientific support and advice, and of S. Landry and V. Stevens for logistic support. Special thanks are extended to Denali National Park Wilderness Centers, LTD (Camp Denali) and its owner, J. Hamm MS for use of the sites, in-kind support, and scientific advice. We thank Dr. R. Monson and three anonymous reviewers for their valuable comments. This research was funded by the American Association of University Women, University of Alaska Fairbanks URSA program, Alaska EPSCoR NSF award #EPS-0701898 and the state of Alaska to KO, and NSF PLR 1107892 to MSB.

REFERENCES

1. Anderson LL, Hu FS, Nelson DM, Petit RJ, Paige KN (2006) Ice-age endurance: DNA evidence of a white spruce refugium in Alaska. P Natl Acad Sci USA 103:12447-12450
2. Barber VA, Juday GP, Finney BP (2000) Reduced growth of Alaskan white spruce in the twentieth century from temperature-induced drought stress. Nature 405(6787):668-673
3. Bret-Harte MS, Shaver GR, Chapin FS (2002) Primary and secondary stem growth in arctic shrubs: implications for community response to environmental change. J Ecol 90(2):251-267
4. Bret-Harte MS, Mack MC, Goldsmith GR, Sloan DB, DeMarco J, Shaver GR, Ray PM, Biesinger Z, Chapin FS (2008) Plant functional types do not predict biomass responses to removal and fertilization in Alaskan tussock tundra. J Ecol 96:713-726
5. Callaway RM (1998) Competition and facilitation on elevation gradients in subalpine forests of the northern Rocky Mountains, USA. Oikos 82:561-573
6. Callaway RM, Brooker RW, Choler P, Kikvidze Z, Lortie CJ, Michalet R, Paolini L, Pugnaire FI, Newingham B, Achehoug ET, Armas C, Kikodze D, Cook BJ (2002) Positive interactions among alpine plants increase with stress. Nature 417:844-848
7. Carter RE (1992) Diagnosis and interpretation of forest stand nutrient status. In: Chappell HN, Weetman GF, Miller RE (eds) Forest fertilization: sustaining and improving nutrition and growth of western forests. College of Forest Resources Contribution #73. University of Washington, Seattle, pp 90–97
8. Cavieres LA, Arroyo MTK, Molina-Montenegro M, Torres C, Peñaloza A (2002) Nurse effect of *Bolax gummifera* (*Apiaceae*) cushion plants in the alpine vegetation of the Chilean Patagonian Andes. J Veg Sci 13:547-554
9. Choler P, Michalet R, Callaway RM (2001) Facilitation and competition on gradients in alpine plant communities. Ecology 82:3295-3308 Publisher Full Text

10. Cortini F, Comeau PG, Bokalo M (2012) Trembling aspen competition and climate effects on white spruce growth in boreal mixtures of Western Canada. Forest Ecol Manag 277:67-73
11. Cortini F, Comeau PG (2008) Evaluation of competitive effects of green alder, willow and other tall shrubs on white spruce and lodgepole pine in Northern Alberta. Forest Ecol Manag 255:82-91
12. Dalen L, Hofgaard A (2005) Differential regional tree line dynamics in the Scandes Mountains. Arct Antarct Alp Res 37:284-296
13. Danby RK, Hik DS (2007) Responses of white spruce (*Picea glauca*) to experimental warming at a subarctic alpine treeline. Glob Change Biol 13:437-451
14. de Lafontaine G, Turgeon J, Payette S (2010) Phylogeography of white spruce (*Picea glauca*) in eastern North America reveals contrasting ecological trajectories. J Biogeogr 37:741-751
15. Díaz S, Hodgson JG, Thompson K, Cabido M, Cornelissen JHC, Jalili A, Montserrat-Martí G, Grime JP, Zarrinkamar F, Asri Y, Band SR, Basconcelo S, Castro-Díez P, Funes G, Hamzehee B, Khoshnevi M, Pérez-Harguindeguy N, Pérez-Rontomé MC, Shirvany FA, Vendramini F, Yazdani S, Abbas-Azimi R, Bogaard A, Boustani S, Charles M, Dehghan M, deTorres-Espuny L, FalczukV, Guerrero-Campo J, Hynd A, Jones G, Kowsary E, Kazemi- Saeed F, Maestro-Martínez M, Romo-Díez A, Shaw S, Siavash B, Villar-Salvador P, Zak MR (2004) The plant traits that drive ecosystems: evidence from three continents. J Veg Sci 15(3):295-304
16. Gervais BR, MacDonald GM (2000) A 403-year record of July temperatures and treeline dynamics of Pinus sylvestris from the Kola Peninsula, northwest Russia. Arct Antarct Alp Res 32:295-302
17. Goldstein G, Brubaker GL, Hinckley T (1985) Water relations of white spruce (*Picea glauca* (Moench) Voss) at treeline in north-central Alaska. Can J For Res 15:1081-1087
18. Grime JP (2001) Plant strategies, vegetation processes, and ecosystem properties. John Wiley & Sons Ltd, Chichester.
19. Grime J, Thompson K, Hunt R, Hodgson JG, Cornelissen JHC, Rorison IH, Hendry GAF, Ashenden TW, Askew AP, Band SR,

Booth RE, Bossard CC, Campbell BD, Cooper JEL, Davison AW, Gupta PL, Hall W, Hand DW, Hannah MA, Hillier SH, Hodkinson DJ, Jalili A, Liu Z, Mackey JML, Matthews N, Mowforth MA, Neal AM, Reader RJ, Reiling K, Ross-Fraser W, Spencer RE, Sutton F, Tasker DE, Thorpe PC, Whitehouse J (1997) Integrated screening validates primary axes of specialisation in plants. Oikos 79(2):259-281

20. Grossnickle SC (2000) Ecophysiology of Northern spruce species: the performance of planted seedlings. NRC Research Press, Ottawa.
21. Hamm J (2007) Recent tree line advance and the influence of shrub and tundra communities on white spruce (Picea glauca) establishment in Denali National Park, Alaska. Master thesis. Department of Environmental Studies, Antioch New England Graduate School, Antioch University, USA
22. Heskel MA, Atkin OK, Turnbull MH, Griffin KL (2013) Bringing the Kok effect to light: a review on the integration of daytime respiration and net ecosystem exchange. Ecosphere 4:art98
23. Hobbie SE, Chapin FS (1998) An experimental test of limits to tree establishment in Arctic tundra. J Ecol 86:449-461
24. Hobbie JE, Hobbie EA, Drossman H, Conte M, Weber JC, Shamhart J, Weinrobe M (2009) Mycorrhizal fungi supply nitrogen to host plants in Arctic tundra and boreal forest: ^{15}N is the key signal. Can J Microbiol 55:84-94
25. IPCC (2008) Synthesis report. In: Core Writing Team, Pachauri RK, Reisinger A (eds) Climate change 2007: Synthesis report. Contribution of working groups I, II and III to the fourth assessment report of the Intergovernmental Panel on Climate Change. Intergovernmental Panel on Climate Change, Geneva
26. Juday GP, Barber V, Duffy P, Linderholm H, Rupp TS, Sparrow S, Vaganov E, Yarie J (2005) Forests, land management,and agriculture. In: ACIA (ed) Arctic Climate Impact Assessment. Cambridge University Press, New York
27. (2009) Regional climate impact: Alaska. Global climate change impacts in the United States. Cambridge University Press, Cambridge.

28. Kok B (1948) A critical consideration of the quantum yield of Chlorella-photosynthesis. Enzymologia 13:1-56
29. Körner C (2012) Treelines will be understood once the functional difference between a tree and a shrub is. Ambio 41:197-206
30. Kullman L (1991) Structural change in a subalpine birch woodland in north Sweden during the past century. J Biogeogr 18:53-62
31. Litton CM, Giardina CP (2008) Below-ground carbon flux and partitioning: global patterns and response to temperature. Funct Eco 22(6):941-954 doi:10.1111/j.1365-2435.2008.01479.x
32. Livingston NJ, Guy RD, Sun ZJ, Ethier GJ (1999) The effects of nitrogen stress on the stable carbon isotope composition, productivity and water use efficiency of white spruce (*Picea glauca* (Moench) Voss) seedlings. Plant Cell Environ 22:281-289
33. Lloyd AH, Fastie CL (2003) Recent changes in treeline forest distribution and structure in interior Alaska. Ecoscience 10(2):176-185
34. Man R, Lieffers VJ (1997) Seasonal photosynthetic responses to light and temperature in white spruce (*Picea glauca*) seedlings planted under an aspen (*Populus tremuloides*) canopy and in the open. Tree Physiol 17:437-444
35. Nienstaedt H, Zasada JC (1990) Picea glauca. In: Burns RM, Honkala BH (eds) Silvics of North America: 1. Conifers; 2. Hardwoods. Agriculture Handbook 654. U.S. Department of Agriculture, Forest Service, Washington, DC
36. Rixen C, Mulder CP (2009) Species removal and experimental warming in a subarctic tundra plant community. Oecologia 161(1):173-186
37. Roland CA, Schmidt JH, Johnstone JF (2013) Climate sensitivity of reproduction in a mast- seeding boreal conifer across its distributional range from lowland to treeline forests. Oecologia: 1–13 doi 10.1007/s00442-013-2821-6
38. Slot M, Wirth C, Schumacher J, Mohren GM, Shibistova O, Lloyd J, Ensminger I (2005) Regeneration patterns in boreal scots pine glades linked to cold-induced photoinhibition. Tree Physiol 25:1139-1150

39. Stitt M, Schulze D (1994) Does Rubisco control the rate of photosynthesis and plant growth? An exercise in molecular ecophysiology. Plant Cell Environ 17:465-487
40. Stueve KM, Isaacs RE, Tyrrell LE, Densmore RV (2011) Spatial variability of biotic and abiotic tree establishment constraints across a treeline ecotone in the Alaska Range. Ecology 92:496-506
41. Sullivan PF, Ellison SB, McNown RW, Brownlee AH, Sveinbjörnsson B (In press) Evidence of soil nutrient availability as the proximate constraint on growth of treeline trees in northwest Alaska. Ecology.
42. Sturm M, Racine C, Tape K (2001) Climate change: increasing shrub abundance in the Arctic. Nature 411:546-547
43. Sveinbjörnsson B (2000) North American and European treelines: external forces and internal processes controlling position. AMBIO: A J Hum Environ 29(7):388-395
44. Szeicz JM, MacDonald GM (1995) Recent white spruce dynamics at the subarctic alpine treeline of northwestern Canada. J Ecol 83:873-885
45. Tape K, Sturm M, Racine C (2006) The evidence for shrub expansion in Northern Alaska and the Pan-Arctic. Glob Change Biol 12:686-702
46. Tilman D (1988) Plant strategies and the structure and dynamics of plant communities. Princeton University Press, Princeton.
47. USDA, ARS, National Genetic Resources Program. Germplasm Resources Information Network - (GRIN) [Online Database]. National Germplasm Resources Laboratory, Beltsville, Maryland. URL: http://www.ars-grin.gov/cgi-bin/npgs/html/taxon.pl?433404 (25 October 2014)
48. Waring RH (1987) Characteristics of trees predisposed to die. Stress causes distinctive changes in photosynthate allocation. BioScience 37(8):569-574
49. Way DA, Sage RF (2008) Thermal acclimation of photosynthesis in black spruce [*Picea mariana*(Mill.) BSP]. Plant Cell Environ 31(9):1250-1262
50. Wilmking M, Juday GP, Barber V, Zald HSJ (2004) Recent climate warming forces contrasting growth responses of white spruce at

treeline in Alaska through temperature thresholds. Glob Change Biol 10:1724-1736

51. Yin HJ, Liu Q, Lai T (2008) Warming effects on growth and physiology in the seedlings of the two conifers *Picea asperata* and *Abies faxoniana* under two contrasting light conditions. Ecol Res 23:459-469
52. Zhang SR, Dang QL (2007) Interactive effects of soil temperature and CO_2 on morphological and biomass traits in seedlings of four boreal tree species. Forest Sci 53:453-46

Chapter 7

The Roots of Future Rice Harvests

Nourollah Ahmadi[1], Alain Audebert[1], Malcolm J Bennett[2], Anthony Bishopp[2], Antonio Costa de Oliveira[3], Brigitte Courtois[1], Abdala Diedhiou[4,5], Anne Diévart1, Pascal Gantet[6,7],Alain Ghesquière[8], Emmanuel Guiderdoni[1], Amelia Henry[9], Yoshiaki Inukai[10], Leon Kochian[11], Laurent Laplaze[5,8], Mikael Lucas[8], Doan Trung Luu[1,2], Baboucarr Manneh[13],Xiaorong Mo[14], Raveendran Muthurajan[15], Christophe Périn[1], Adam Price[16], Sabariappan Robin[15], Hervé Sentenac[12], Bassirou Sine[17,5], Yusaku Uga[1,8], Anne Aliénor Véry[1,2],Matthias Wissuwa[19], Ping Wu[14], and Jian Xu[20]

[1]CIRAD, UMR AGAP, Montpellier Cedex 5, 34398, France

[2]Centre for Plant Integrative Biology, University of Nottingham, Loughborough LE12 5RD, UK

[3]UFPel, Plant Genomics and Breeding Center, Pelotas, Brazil

[4]Université Cheikh Anta Diop (UCAD), Département de Biologie Végétale, Laboratoire Commun de Microbiologie IRD/ISRA/UCAD, Centre de Recherche de Bel Air - BP 1386 CP 18524, Dakar, Sénégal

[5]Laboratoire Mixte International Adaptation des Plantes et microorganismes associés aux Stress Environnementaux (LAPSE), Centre de Recherche de Bel Air - BP 1386 CP 18524, Dakar, Sénégal

[6]Université Montpellier 2, UMR DIADE, Montpellier, France

[7]IRD, LMI RICE, USTH, Agronomical Genetics Institute, Hanoi, Vietnam

[8]IRD, UMR DIADE, Montpellier, France

[9]IRRI, Los Baños, Philippines

[10]International Cooperation Center for Agricultural Education (ICCAE), Nagoya University, Furo-cho, Chikusa 464-8601, Nagoya, Japan

[11]Robert W. Holley Center for Agriculture and Health, USDA-ARS and Department of Plant Biology, Cornell University, Ithaca 14853, NY, USA

[12]Biochimie et Physiologie Moléculaire des Plantes, Institut de Biologie Intégrative des Plantes, UMR 5004 CNRS/386 INRA/Montpellier SupAgro/Université Montpellier 2, F-34060, Montpellier Cedex 2, France

[13]Africa Rice Center, AfricaRice Sahel Regional Station, B.P. 96, St Louis, Senegal

[14]State Key Laboratory of Plant Physiology and Biochemistry, College of Life Science, Zhejiang University, Hangzhou 310058, China

[15]Tamil Nadu Agricultural University, Coimbatore, 641003, India

[16]University of Aberdeen, Aberdeen AB24 3UU, UK

[17]ISRA, CERAAS, Thiès, Senegal

[18]National Institute of Agrobiological Sciences (NIAS), 2-1-2 Kannondai, Tsukuba 305-8602, Ibaraki, Japan

[19]Japan International Research Center for Agricultural Sciences (JIRCAS), 1-1 Ohwashi, Tsukuba 305-8686, Japan

[20]Department of Biological Sciences and NUS Centre for BioImaging Sciences, Faculty of Science, National University of Singapore, Singapore 117543, Singapore

ABSTRACT

Rice production faces the challenge to be enhanced by 50% by year 2030 to meet the growth of the population in rice-eating countries. Whereas yield of cereal crops tend to reach plateaus and a yield is likely to be deeply affected by climate instability and resource scarcity in the coming decades, building rice cultivars harboring root systems that can maintain performance by capturing water and nutrient resources unevenly distributed is a major breeding target. Taking advantage of gathering a community of rice root biologists in a Global Rice Science Partnership workshop held in Montpellier, France, we present here the recent progresses accomplished in this area and focal points where an international network of laboratories should direct their efforts.

REVIEW

The root system has the crucial role of extracting nutrients and water through a complex interplay with soil biogeochemical properties, and of maintaining these functions under a wide range of stress scenarios to ensure plant survival and reproduction. This role is made even more important due to increasing climate instability and limitation of fertilizers and irrigation in cropping systems. Determining the precise contribution of root traits to final grain yield under these scenarios and breeding cultivars harboring root systems adapted to stress profiles prevalent in representative target soil environments are therefore a priority in the plant breeding research agenda. Long neglected, the biology and ecology of roots - the "hidden half" - have recently attracted an increasing number of research groups and disciplines ranging from genetics, molecular biology, cell biology, physiology, microbiology, engineering, and biomathematics. This attraction has been exemplified by the high attendance of the International Root Research Society meetings (http://www.rootresearch.org). Demonstrations at the field level that root anatomical and architectural traits can considerably enhance cereal crop yield stability under limited water and nutrient resources (Lopes and Reynolds [2010], York et al. [2013], Lynch et al. [2014]), as well as the development of high-throughput root phenotyping systems (Clark et al. [2013], Topp et al.[2013]) have also recently contributed to this interest. Furthermore, more groups working in root development in the model plant *Arabidopsis* are now turning to crops, such as rice (*Oryza sativa*L.), the model cereal.

Addressing issues of water and nutrient limitations in rice, the staple food for more than half of mankind, is of particular relevance since rice is cultivated under the widest range of agro-systems among cereals - from irrigated anaerobic conditions to upland aerobic conditions, including flood-prone and drought-prone environments. Breeding for high-yielding irrigated rice has favored shallow root system phenotypes that easily capture resources in the topsoil layer whereas breeding for low input upland conditions has, conversely, favored deep root systems with strong foraging capacity and nutrient extraction ability from poor soil layers. In addition to these divergent root architectures that have been selected for among rice agro-ecosystems, the increasingly fluctuating levels of stress that is projected to occur across rice growing systems may necessitate more diverse root architectures and better adaptability to uneven distribution of soil resources. The diversity of rice for adaptation to different water regimes and soil conditions has favored the identification of genes and quantitative trait loci (QTLs) underlying root development and resource uptake traits. In parallel, methodological progress in the development of experimental set ups that may mimic uneven resource distribution, has led to the establishment of diverse and complementary phenotyping platforms for screening for root traits in diversity panels, mapping populations and mutant resources. Though their relevance to field soil conditions has to be considered with caution, such platforms are instrumental for feeding, then calibrating and validating, dynamic models of root architecture in response to resource scarcity and distribution (reviewed in Hill et al. [2013]). Modeling is likely to be of invaluable assistance for breeding ideotypes adapted to specific soil environments and stress scenarios as millions of simulations can be run encompassing a wide range of environmental conditions (Postma et al. [2014]).

To review the progress in these areas and delineate future priorities where research efforts on rice roots should focus, a workshop gathering 40 participants was organized in the Agropolis campus, Montpellier France, from May 15-16th 2014, in the framework of the Global Rice Science Partnership (GRiSP). The group has first focused its debates on advances in root development and ideotype breeding. As a major output of the discussions, we recommend research areas on which emphasis should be placed that constitute a priority set of core actions of a rice root research collaborative network.

Accelerate the Discovery of Genes Underlying Root Structural and Functional Traits

During the past two years, the rice research community has seen major breakthroughs in the identification of genes behind root response to stress; *DRO1*, a root depth QTL (Uga et al., [2013]) and *PUP1* a root

QTL for phosphorus uptake (Gamuyao et al., [2012]) were cloned, and fine mapping of several other rice root/stress response QTLs is ongoing (*DRO2, DRO3, CHR9*; Y. Uga and N. Ahmadi et al. unpublished work). Recent advances in MetaQTL analyses and genome wide association mapping approaches (GWAS) can reduce QTL confidence intervals and can help to improve the precision of the number and position of detected QTLs (Courtois et al. [2009], [2013]). These approaches are time consuming, and shortening the timeframe for QTL characterization is clearly a priority for the future.

The most effective strategy for cloning these QTLs appears to be the development of Near-Isogenic Lines (NILs) by marker-assisted selection. These NILs are useful material for fine mapping and QTL cloning but also for breeding and for detailed physiological studies. The NIL strategy combines breeding, exploitation of genetic diversity and positional cloning (Motte et al.[2014]). The work around root architecture and phosphate uptake also illustrates the importance of identifying cultivars that have special behavior (e.g. highly efficient P uptake despite a poor root system) to include rare alleles that will be valuable for breeding. Rice CSSL (Chromosome Segment Substitution Lines) populations, are valuable/natural starting points for QTL mapping and subsequent QTL cloning. Alternatively, developing new populations, especially NILs targeting QTLs of large effect is still a necessary step for its positional cloning.

The rice root system displays a complex structure composed of several tissues and different root types (Rebouillat et al. [2009]). The function of an increasing range of genes involved in rice root development has been deciphered these last 5 years (reviewed in Coudert et al. [2010]; Orman-Ligeza et al. [2013]). Root stress responses are clearly influenced by their cell identity and coordination of cell-specific responses that are currently poorly understood. All of the root tissues, including root cap (Wang et al. [2014]), could develop anatomical and molecular adaptations related to their diverse roles in root growth and function under normal and stress conditions. Exploring root stress responses and adaptations at a cell or tissue specific level by characterization of gene networks involved in these processes is an expanding research field (Gifford et al.[2008]). The use of gene regulatory network inference tools to identify key regulators of root adaptive responses need also to be used in rice. Such an approach should yield candidate master regulators of root responses to stress and is complementary to genetic approaches. The identification of common cis-acting regulatory elements can aid to gather co-expressed genes under root development and stress responses (Pegoraro et al. [2013]).

The validation of QTLs and reverse genetics often involves use of *A. thaliana* data. There is a need to improve and speed up annotation transfer between *Arabidopsis* and rice. Identification of orthologs is probably the most straightforward way. Expression data or synteny can be also used to find the right ortholog(s) between species. Databases of orthologs already exist (Goodstein et al. [2012]; Rouard et al. [2011]) as well as new tools for ortholog prediction (Ostlund et al. [2010]; Schreiber and Sonnhammer, [2013]) but the challenge is to build a simple and functional platform for biologists and breeders.

Functional validation of individual genes associated with root development through knockout or gain-of-function mutants still holds merit. Since the multiple genes of roots are found to have spatial and temporal differences in their expression and function, the search for the trait differences, genes and their function need to be expanded with large set of mutant resources (Wei et al. [2013] for a review). This includes the development of large saturation mutagenized populations, virtually exploring any amino acid variation along the encoded proteins. New gene targeting technologies, mainly TALEN and CRISPR are revolutionizing functional analysis of genes and should be critical to accelerate the characterization of genes of agronomic interest (Gaj et al.[2013]; Miao et al. [2013]). The establishment of an International repository containing CRISPR directed against all rice genes to generate knock-out mutants would be a helpful resource for speeding up gene discovery and validation. Another possibility would be to produce CRISPR in pairs targeting 5′ and 3′ end of each rice gene. This resource would be useful for allele replacement in any rice gene and to validate any QTL effect by introducing only the causative mutation in other genetic backgrounds. Finally, the CRISPR technology allows targeting multiple genes at once in a single transformation event and so opens a simple way to test QTL/gene interactions.

Establish High Throughput Phenotyping Platforms with Relevance to Field Conditions

GWAS is promoted by the recent progresses in re-sequencing rice genetic resources (CAAS BGI IRRI, [2014]) and reverse and direct genetics approaches are facilitated by the creation of large and diverse mutant collections (Wei et al. [2013]). Dedicated high-throughput phenotyping methods coupled with automated data analysis systems are necessary to fully exploit these genetic resources to identify the genetic determinants of root (adaptive) architecture.

The choice of a phenotyping method always results from a compromise between relevance, precision, throughput, automation and cost to address a specific scientific question. Some key root QTLs have been discovered indirectly without the necessity to screen roots. This is the case of PUP1 that was discovered merely on the basis of the capacity of plants to maintain yield when grown in phosphorus-depleted soils. It is only *a posteriori* that *PSTOL1*, the isolated gene underlying *PUP1* QTL, proved to control an adaptive development response of the root system in low-phosphate soil condition promoting root branching (Gamuyao et al. [2012]). Other important root genetic determinants governing rice adaptation to soil conditions could be found by this indirect way, but it will be probably limited. Most studies performed in the last decades were based on direct observation of the root system *per se*. The use of hydroponics-, soil pots- or soil columns-based systems and the collection of traits such as maximum root length, root thickness, root number, mass of roots at different depths, or root to shoot ratio allowed the identification of several QTLs (Courtois et al. [2009]). Some of these traits have already been integrated into breeding programs, e.g. to increase water deficit tolerance by increasing deep rooting (Shen et al.[2007]). Another soil-based and indirect root screening system uses introduction of herbicide at a deep soil layer (Khalfaoui and Havard, [1993]). In this case the plant starts to die when its root system reaches the herbicide layer. This reflects the rapidity with which the root develops in depth which is a combination of root gravity perception (or root angle) and growth rate (Al-Shugeairy et al. [2014]). Another system, called *shovelomics* uses the partial digging up of the root system under field conditions, through a rapid (8 minutes per plant) phenotyping of some parameters such as root biomass in top soil and the total number of crown roots (Trachsel et al. [2011]). This method gives partial data and does not estimate root depth for example. All these soil-based system have proven their usefulness to discover some major root genes and QTLs. Nevertheless, these systems are destructive and limit fine and dynamic study of root development.

A new alternative experimental set up allows the *in situ* capture of the whole root system in soil pots or columns using X-ray computed tomography (Tracy et al. [2010]; Bao et al. [2014]). This is a non-destructive process, and plants can be imaged at multiple stages of their lifecycle. Though time needed for image acquisition and segmentation currently limits its use to a restricted set of genotypes, it is very promising in term of discovery of genes involved in root developmental adaptive response and deeper characterization of varieties identified through other screens. Other systems focus on early developmental stage and use transparent media that are considered to mimic soil. For example the Rhizoscope, a two-dimensional hydroponic-based system made of plexiglass sandwiches filled with glass beads that mimic soil resistance, allows root system growth up to 30 days after sowing. Root growth can be examined and fine and quantitative parameters can be collected after glass beads removal, root imaging and tissue harvesting. This system has been used to phenotype the roots of a *japonica* rice core collection (Courtois et al. [2013]). *In vitro*agar-medium 3-D based systems have also been developed (Iyer-Pascuzzi et al. [2010]; Clark et al. [2011]). Other systems are restricted to very early developmental stages such as the rolled paper systems where the medium is supplied by capillarity in the support paper and that give access to a bi-dimensional view of the early root system still anchored on the paper after opening (Woll et al. [2005]). These systems allows the following of the dynamics of early root development in a high throughput manner and will be probably very useful to screen the root architecture of rice core or mutant collections and to study the adaptive response of the root system to different nutrients in soil.

Understanding the detailed components of rice root growth, in terms of cell differentiation, anatomy, ion fluxes, and hydraulic conductance requires phenotyping systems that are operational at a finer scale in addition to those characterizing whole root-system architecture. Hydraulic properties of segments of specific root zones were investigated in several cultivars of rice (Ranathunge et al. [2003]). This approach allows a fine description of the water transport capacity along the root, and, in combination with drugs inhibiting the aquaporin intrinsic activity, allows one to decipher the respective contributions of apoplast and cell-to-cell paths (Ranathunge et al.[2004]). The diversity of rice root anatomy is also starting to be systematically explored by image analysis of histological sections. This is promising because the radial tissue differentiation is involved in the capacity of rice to better stand toxicity of minerals such as aluminum and water deficit (Huang et al. [2009]; Henry et al. [2012]). Both tissue section preparation and image analysis can be automated in the future (Périn C., Henry A. pers com). One potential method of investigating 3D cellular structures is the technique of laser ablation tomography that has recently been developed for maize roots (http://plantscience.psu.edu/research/labs/roots/sites/laserfacility). In this technology the surface of a root is alternatively ablated with a pulsed laser and photographed with a dSLR camera. This has now been used in a variety of plant species and is likely to provide a powerful tool for rice research. Four dimensional-reconstruction of a growing rice root tip with cell lineage tracking has also proven feasible (Fernandez et al. [2010]). Recent*Arabidopsis* work has demonstrated that this level of phenotyping allows the GWAS identification of novel important genes involved in root development (Meijon et al. [2014]).

Open and high-throughput root phenotyping platforms that can address root growth in realistic conditions need to be developed in the future. Another important bottleneck is the extraction of meaningful quantitative data from the large amount of images generated by high throughput phenotyping platforms. Several image analysis software products have been developed to assist the root phenotyping platforms (French et al. [2009]; Galkovskyi et al. [2012]; Burton et al.[2012], Clark et al. [2013]; Pound et al. [2013]; Topp et al. [2013]) but so far the image analysis still requires a time-consuming, expert curation step. Although the choice of a phenotyping method is always a compromise between the available means and the scientific question, some reference protocols and parameters as well as priority issue questions should be defined and adopted by the rice community as it has been done for other models (De Smet et al. [2012]). This would allow the rice scientific community to exploit common genetic resources (sequenced genotypes, highly recombinant populations, mutant collections) and to share data such as raw images that could be exploited for different phenotypic traits by different groups. This could apply for early developmental stage with simple standardized culture systems to investigate constitutive and adaptive root development and also for root anatomy. A common repository for these phenotyping data would also pave the way for analyses of genotype x environment interactions, meta-analyses and comparison between different experiments.

Delineate Root Ideotypes Adapted to Diverse Target Soil Environments

For future crops, it is expected that multiple root characteristics should be improved, including those that confer improved water use efficiency and nutrient acquisition (Pennisi [2010]). In the case of rice, one strategy has been to identify genes controlling such interesting root traits, which relies on the direct identification of genetic regions conferring tolerance to a given stress using a genetic screen, GWAS, or QTL analyses (e.g. *PUP1*; Gamuyao et al. [2012]). An alternative is to try to define *a priori* ideotypes of more efficient root systems from our knowledge of soil sciences and plant physiology. Root ideotypes that have been suggested to improve rice performance under drought include increased root growth at depth (Nicou et al. [1970]; O'Toole and Chang [1978]) and coarse axial roots to increase the ability to penetrate hardpans (Hasegawa et al. [1985]). The hypothesis that deep-rooted rice would be more resistant to drought than shallow rooted ones prompted the identification of a gene (*DRO1*) controlling nodal root growth angle in rice, and NIL having the functional allele of *DRO1* showed improved yield under upland drought conditions (Uga et al. [2013]). Similarly, deep rooting has been hypothesized as a trait that could allow for exploitation of soil nutrients at deeper soil layers thus improving mineral uptake and potentially increase yields (Obara et al. [2010]). NILs having the Kasalath allele at QTL qRL6.1 had longer seminal roots than those with the Koshihikari allele under different nitrogen fertilizer concentrations with NH4+ as the source of nitrogen. These examples demonstrate the potential value of ideotype-based approaches. However, as our knowledge of yield determination and response to stress increases, the complexity of the mechanisms involved in rice root response to stress and the diversity of environmental conditions in rice agro-ecosystems make the definition of ideotypes more difficult. Both the approaches of root ideotype-based screening and selection for grain yield are not mutually exclusive and should be conducted in parallel. They would in fact sustain each other; ideotype-based approaches may improve stress response and allow the testing of the ideotypes, and yield-based screens will provide better knowledge of the physiology of stress tolerance that will help refine the ideotypes. Such refined ideotypes should lead our formation of strategies to design new genetic screens based on a better knowledge of the integrated stress responses (physiology, development and plant-microbe interactions) and might help identify interesting alleles that might not be identified otherwise. For example, an experimental approach based on better knowledge of the physiology of water transport/acquisition has the potential to provide more detailed insights to mechanisms behind drought tolerance than a screen based on grain yield alone. Two major factors need to be addressed in order to optimize the benefit of such an ideotype-based approach: the understanding of environmental conditions and the integration of the complex factors behind stress response.

To optimize the outputs of root biology for crop improvement, it is first important to better define the environmental stresses that are faced by rice culture worldwide. This points to the need for better characterization of the actual (on-farm) target environments in which crops are grown (including soil physical properties, as in the approaches reported by Inthavong et al. [2011] or Haefele et al. [2014]), and the need for good estimates of future scenarios for environmental conditions faced by rice growers. This will help define the major limiting factors that can be addressed through root ideotype breeding, and to target the most important varieties specific to target each type of environmental conditions to maintain food production in the future (i.e., upland and rainfed lowland drought stress, or fluctuating soil moisture conditions).

Plant responses to environmental stresses are complex and involve multiple levels of integration. Root architecture has been the focus of most proposed ideotypes in the past but anatomical and physiological traits are likely to play equally key roles. Some research on water transport ideotypes has been conducted through studies on the effect of xylem vessel diameter (Yambao et al. [1992]; Henry et al. [2012]) but there is not a consensus in the literature as to whether larger or smaller xylem vessel diameters are ideal. Establishing, a definite answer to these issues will be important, although it should be borne in mind that different xylem diameters/configurations may be optimal for different environments. Given the predicted increasing fluctuations in rice-growing climates, genotypes exhibiting a high degree of root growth plasticity will likely increase in importance (Kano et al. [2011]). Phenotypic effects of beneficial microbes (fungi and bacteria) on plant root system include primary and lateral root length, number and positioning of lateral roots, and overall root growth (Dodd and Ruiz-Lozano [2012], Sukumar et al. [2013]). For instance, it has been shown in rice that the arbuscular mycorrhizal (AM) fungus, *Rhizophagus irregularis*, stimulates the formation of fine and large lateral roots (Gutjahr et al. [2009]). On the other hand, AM fungi form hyphal networks which operate as functional extensions of the plant root system (Smith & Read [2008]). Hence, it appears that beneficial microbes can be considered as an integral part of the root system, from which they significantly contribute to nutrient acquisition and host tolerance to occasional or prolonged abiotic and biotic stress. Our perspective is that, in addition to root architecture, future work on ideotypes should integrate functional parameters (e.g. water and ion uptake per length of root and the respective contributions of apoplast and cell-to-cell paths), plasticity, and plant/soil microbe interactions which can alter host root development and functioning.

However, as our understanding increases, the concept of root ideotype is less clearly defined. It is therefore important to exploit the power of modeling to integrate these different factors and be able to make predictions about the traits that may be beneficial. Modeling forces us to examine many parameters related to a specific process and allows us to test the sensitivity of the model to each of these. This can inform about potential rate-limiting processes that govern complex traits such as water or nutrient acquisition and guide the subsequent identification of associated QTLs. Several structural functional root models have been published (e.g. SimRoot, Lynch et al. [1997] and R-SWMS, Dunbabin et al. [2013]) as well as other soil models (Reviewed in Hill et al. [2013]). Root/soil interaction models will be important in simulating the dynamic response and efficiency of different rice ideotypes in different soil conditions. One considerable advantage of such models is that they allow us to investigate the cost-benefit relationship between various traits in a wide variety of environments. For example, producing longer roots can protect against drought but it comes at a cost of carbon that could be used elsewhere, and deep roots may be less efficient in foraging for nitrogen. Structural-functional plant models based on an economy of resources allow us to identify potential combinations of traits that can increase plant growth under a wide variety of environmental conditions, including water or nutrient poor conditions. These predictions could drive breeders or genetic engineers towards producing plants with these combinations of traits, so that the models can be tested experimentally. Again, characterization of the environment will be critical for this analysis, and having a geographically distributed network of rice root researchers with knowledge of different soils is an enormous asset. Future models need to take into account the soil structure as well as nutrient/water availability, and incorporate both root architecture and anatomy within the model. These tools will allow the search for more difficult to study root traits that are linked to adaptability, rather than fixed characters for a given environment. This effort should be community-based and where possible should use open access platforms (such as OpenAlea; Pradal et al. [2008]) to guarantee adoption and the continuity of curation of the developed models.

In summary, we think that in addition to the approach of screening for stress tolerance among diverse rice genotypes (including wild relatives), there are many benefits to be gained from developing an approach to root biology based on the integration of our knowledge of the physiology of plant response to stress to define ideotypes that can form the basis of new more specific genetic screens. For instance, our perspective is that future work on ideotypes should integrate functional parameters (i.e. water and nutrient uptake per length of root,). Future research is necessary to conceptualize new integrative root ideotypes that will enable the definition of new experimental screening approaches to contribute to rice improvement strategies. A more complex, but necessary step would be the coupling of the roots development modelling approaches and tools with the ones of the whole plant modeling in order to reason the ideotype at the whole plant level.

Breed for Improved Root Systems of Future Rice Varieties

Several hundred QTLs for root traits (mainly structural and a few functional) have been identified and compiled for the identification of meta-QTLs (Courtois et al. [2009]; Khowaja et al. [2009]). Association mapping, which improves further the resolution in QTL position, has already been initiated for rice root traits on broad-based panels (Famoso et al. [2011]; Courtois et al. [2013]; Wissuva et al. unpublished work) and a core collection of local varieties from Vietnam (Gantet et al. unpublished work). One of these QTLs, *DRO1,* was recently cloned and its function characterized (Uga et al. [2013]). More recently, the characterization of QTLs with large effect involved in phosphorus uptake (Gamuyao et al. [2012]) and in yield under drought (Kumar et al.[2013]; Swamy et al. [2013]) revealed direct relationship with or contribution from root traits. Efforts for cloning a number of other root QTLs are ongoing at NIAS (QTLs for stele size and surface rooting), JIRCAS (QTLs for P uptake), Nagoya University (QTL for root growth in response to soil moisture fluctuations and root growth plasticity) IRRI (QTLs for yield under drought), CIRAD (QTLs for root length) and Tamil Nadu University (QTLs for root penetration ability). Thus a large number root QTLs or genes promising for improving rice performances in water and nutrient uptake are now available for breeders.

Efforts in marker assisted introgression of root QTLs from landraces into elite material go back to 2001 (Shen et al. 2001) with mixed effects on yield in target environment in some cases (Steele et al. 2007). positive effects in some other cases (Gamuyao et al. [2012]; Uga et al. [2013]; Steele et al. [2013]). An important next step for practical use of the large number of root QTLs within breeding programs would be their characterization for QTL x environment and QTL x genetic background interactions, using grain yield as the selection criterion. Indeed, root-related QTLs/genes may interact with the genetic background into which they are introgressed and with the environment to which they are targeted. Thus, it will be necessary to assess the effect of individual or combination of QTLs in the largest possible number of relevant environments and genetic backgrounds, the latest taking into account the structuration of *O. sativa* into four majors groups (*indica, japonica, aus* and *aromatic*). The systematic implementation of such an approach (all combinations of genetic backgrounds and environments) would necessitate a too large set of material to be developed. Simulation modeling could be useful to help discard non-functional trait/QTL combinations at early stages in the development of improved rice genotypes. Subsequently, the results of the yield evaluation together with existing QTL data could be consolidated to develop a breeder's chip carrying markers associated with the main QTLs and candidate genes.

We expect that as more QTLs/target genes are identified and as genetic and genomic approaches converge, a radical change in methodology to rapidly deliver new rice varieties will be necessary. Root architecture is one trait among others for breeders, and it will be necessary to concurrently manipulate or introgress in addition to genes related to root architecture, those for disease resistance, multiple abiotic stress tolerance, grain quality, etc. To deal with such complexity, an emerging solution is genomic selection using a multi-trait index (Meuwissen et al. [2001]). The methodology is being adapted to plants and shows interesting promises (Jannink et al. [2010]; Lorenz et al. [2011]). However the application of genomic selection will require making parallel progress in high throughput phenotyping to calibrate predicting models, in extraction of DNA of the right quality for a high number of plants in a short time, and decreasing even more the cost of genotyping.

To take full advantage of the potential of genomic selection, the process of recombination of multiple parents will also have to be rationalized and accelerated. The best solution to accumulate favorable alleles at multiple genes in a unique background is probably not to try to do it in one round. Since breeding is a numbers game, this would require populations that are too large. Theory demonstrated that a two or more step procedure is much more efficient although slightly slower (Kervella et al. [1993]). Recurrent selection with a male sterility gene to facilitate intercrossing will be a way to go (Grenier et al. unpublished work). As part of the favorable alleles may come from exotic material, the issue of introgression of the genes into an advanced background without disturbing the whole genetic balance has to be solved. This may mean using bridges between wild species or landraces and the elite gene pools. These approaches should help breeders in taking further advantage of the genetic diversity of root traits for improving rice for future harvests.

CONCLUSIONS

The Roots of Future Rice Harvests

In summary, we identified the following priority targets for rice root research during the workshop:

Have a better definition of the environmental stresses that are faced by rice crop worldwide and a better characterization of the actual (on-farm) target environments. For this and to challenge ideotype models, we must mobilize a geographically distributed network of rice root researchers with knowledge of different soils.

Create new populations and new NILs for root QTL cloning

Explore genes and alleles in wild relatives and landraces without neglecting the modern gene pools that can probably also provide interesting features that will be easier to transfer to cultivated rice.

Challenge the effect of the QTLs in several genetic backgrounds taking into account the strong species structuration, across ecosystems and then within each ecosystem. QTL expression has to be analyzed in the largest possible number of relevant situations.

Gain detailed knowledge on the roles that genes corresponding to QTLs play in root and plant physiology (i.e. mineral and water nutrition). Organize/stimulate the exchange of information between research groups aiming at identifying QTLs/having identified QTLs and groups likely to be interested by the investigation of the roles that the corresponding QTL play in root and plant physiology.

Develop tools for ortholog prediction between rice, *Arabidopsis* and crops to accelerate candidate validation and information transfer

Integrate genome editing and allele replacement tools in root QTL validation

Study how roots regulate stress responses at cell/tissue levels through cell specific *omics*approaches and cell-specific gene expression.

Better understand the epigenetic regulation of roots and root development under stressful conditions

Prepare for a capacity of accumulating favorable alleles at multiple genes in a unique background, notably through pyramiding and recurrent selection and establishing intermediary bridges for those residing in exotic germplasm.

Accelerate the development of dedicated high throughput phenotyping systems coupled with automated data analysis

Adopt joint reference protocols and parameters as well as priority issue questions allowing a common and shared effort of harnessing of genetic resources. Create a common phenotyping data repository for comparisons and meta-analyses.

Promote the approaches coupling complementary root ideotype-based screening and selection for grain yield, each approach sustaining the other.

Design new genetic screening methods based on a better knowledge of the integrated stress responses. Integrate functional parameters (e.g.. water uptake per length of root), plasticity, and plant/soil microbe interactions.

Intensify the efforts towards dynamic root/soil interaction modeling to integrate these different factors and be able to make predictions about the traits that may be beneficial under a variety of environmental conditions.

The root system being less accessible and more complex than other agronomical traits, achieving the ambitious goal of breeding the roots of the future requires a coordinated effort and joint resources. The GRiSP workshop held in Montpellier has confirmed that an expanding community, including groups previously using *Arabidopsis* as a model, is at work to tackle the characterization of the structure and function of rice roots and to integrate root traits for breeding improved varieties. This first workshop had no ambition to gather the whole community and tackle all the complex aspects of rice root research. Participants were conscious that many other groups targeting complementary aspects of root biology and ecology research in rice do exist and can join their efforts to this community. Recent advances have been accomplished in the genetic and molecular control of root development, root system phenotyping, imaging and modeling as well as in genome-wide marker assisted breeding and access to nucleotide variation in genetic resources. Altogether, this makes a unique conjunction for a community to build a promising and exciting future in rice root research. In that aim we call for a better coordination of research and breeding efforts through the creation of a strong comprehensive rice root research network as well as for appropriate support from national and international funding bodies. These efforts are both worthwhile and timely since roots are certainly one of the key traits which will help us to adjust crops to the predicted increasing climate instability and resource scarcity with the objective of feeding 9 billion people by year 2050.

AUTHORS' CONTRIBUTIONS

NA, AA, AB, BC, ACO, AD, AD, PG, AH,YI, LL, ML, DTL, BM, RM, XM, LP, AP, SR, YU, AAV, MW, JX wrote the first draft. MJB, AG, LK, HS, BS, PW added valuable comments. EG completed the manuscript writing and did the final polishing. All authors read and approved the final manuscript.

ACKNOWLEDGEMENTS

The authors thank the Global Rice Science Partnership and Agropolis Fondation (Special grant n° 1400–009 and Rhizopolis grant n° 1001–005) benefiting from a national ANR "Investissement d'Avenir" grant ANR-10-LABX-001-01) for supporting the workshop. They acknowledge the assistance of Nathalie Pivot, Cirad and Véronique Rafin, INRA in workshop organization. The root research at Cirad and University of Aberdeen is supported by the European Grant (FP7/2007-2013) under grant agreement n° 289300.27 EURoot "Enhancing resource Uptake from ROOTs under stress in cereal crops". Research at IRRI is supported by the Generation Challenge Program and the Bill and Melinda Gates Foundation. J.X. is supported by the AcRF Tier 2 grant (MOE2009-T2-1-060) from the Ministry of Education of Singapore and National Research Foundation Singapore under its Competitive Research Programme (CRP Award No. NRF2010NRF-CRP002-018). Doan Trung Luu is supported by the EU Marie Curie International Outgoing Fellowship 'ORYZAQUA – Cell Biology of Rice Aquaporins' (PIOF-GA-2011-300150). AP acknowledges the Generation Challenge Programme funded project "Targeting drought avoidance root traits to enhance rice productivity under water limited environments". Financial support for A.G. Diedhiou was provided by the Université Cheikh Anta Diop (UCAD, VE12/13, CpVIII-Ar4) and GRISP.

*This paper is dedicated to the late memory of Pr Ping Wu who passed away in a tragic car accident on June 12th, 2014.

REFERENCES

1. Al-Shugeairy Z, Islam MS, Shrestha R, Al-Ogaidi F, Norton GJ, Price AH (2014) High throughput screening of rooting depth in rice using buried herbicide. Ann Appl Biol 165:96-107
2. Bao Y, Aggarwal P, Robbins NE II, Sturrock CJ, Thompson MC, Tan HQ, Tham C, Rodriguez PL, Vernoux T, Mooney SJ, Bennett MJ, Dinneny JR (2014) Plant roots employ a patterning mechanism to position lateral root branches towards available water. Proc Natl Acad Sci 111(25):9319-9324
3. Burton AL, Williams M, Lynch JP, Brown KM (2012) RootScan: Software for high-throughput analysis of root anatomical traits. Plant Soil 357:189-203
4. CAAS BGI IRRI, c (2014) The 3,000 rice genomes project. Giga Sci 3:7
5. Clark RT, MacCurdy RB, Jung JK, Shaff JE, McCouch SR, Aneshansley DJ, Kochian LV (2011) Three-dimensional root phenotyping with a novel imaging and software platform. Plant Physiol 156:455-465
6. Clark RT, Famoso AN, Zhao K, Shaff JE, Craft EJ, Bustamante CD, McCouch SR, Aneshansley DJ, Kochian LV (2013) High-throughput two-dimensional root system phenotyping platform facilitates genetic analysis of root growth and development. Plant Cell Environ 36:454-466
7. Coudert Y, Perin C, Courtois B, Khong NG, Gantet P (2010) Genetic control of root development in rice, the model cereal. Trends Plant Sci 15:219-226
8. Courtois B, Ahmadi N, Khowaja F, Price A, Rami JF, Frouin J, Hamelin C, Ruiz M (2009) Rice root genetic architecture: meta-analysis from a drought QTL database. Rice 2:115-128
9. Courtois B, Audebert A, Dardou A, Roques S, Ghneim-Herrera T, Droc G, Frouin J, Rouan L, Goze E, Kilian A, Ahmadi N, Dingkuhn M (2013) Genome-wide association mapping of root traits in a japonica rice panel. PLoS One 8:e78037

10. De Smet I, White PJ, Bengough AG, Dupuy L, Parizot B, Casimiro I, Heidstra R, Laskowski M, Lepetit M, Hochholdinger F, Draye X, Broadley M, Péret B, Zhang H, Hammond JP, Fukaki H, Mooney S, Lynch JP, Nacry P, Schurr U, Laplaze L, Benfey P, Beeckman T, Bennett M (2012) Analyzing lateral root development: how to move forward. Plant Cell 24:15-20
11. Dodd IC, Ruiz-Lozano JM (2012) Microbial enhancement of crop resource use efficiency. Curr Opin Biotechnol 23(2):236-242
12. Dunbabin VM, Postma J, Schnepf A, Pagès L, Javaux M, Wu L, Leitner D, Chen YL, Rengel Z, Diggle AJ (2013) Modelling root–soil interactions using three–dimensional models of root growth, architecture and function. Plant Soil 372(1–2):93-124
13. Famoso AN, Zhao K, Clark RT, Tung CW, Wright MH, Bustamante C, Kochian LV, McCouch SR (2011) Genetic architecture of aluminum tolerance in rice (Oryza sativa) determined through genome-wide association analysis and QTL mapping. PLoS Genet 7(8):e1002221
14. Fernandez R, Das P, Mirabet V, Moscardi E, Traas J, Verdeil JL, Malandain G, Godin C (2010) Imaging plant growth in 4D: robust tissue reconstruction and lineaging at cell resolution. Nat Methods 7:547-553
15. French A, Ubeda-Tomás S, Holman TJ, Bennett MJ, Pridmore T (2009) High-throughput quantification of root growth using a novel image-analysis tool. Plant Physiol 150:1784-1795
16. Gaj T, Gersbach CA (2013) ZFN, TALEN, and CRISPR/Cas-based methods for genome engineering. Trends Biotech 31(7):397-405
17. Galkovskyi T, Mileyko Y, Bucksch A, Moore B, Symonova O, Price CA, Topp CN, Iyer-Pascuzzi AS, Zurek PR, Fang S, Harer J, Benfey PN, Weitz JS (2012) GiA Roots: software for the high throughput analysis of plant root system architecture. BMC Plant Biol 12:116
18. Gamuyao R, Chin JH, Pariasca-Tanaka J, Pesaresi P, Catausan S, Dalid C, Slamet-Loedin I, Tecson-Mendoza EM, Wissuwa M, Heuer S (2012) The protein kinase Pstol1 from traditional rice confers tolerance of phosphorus deficiency. Nature 488:535-539
19. Gifford ML, Dean A, Gutierrez RA, Coruzzi GM, Birnbaum KD (2008) Cell-specific nitrogen responses mediate developmental plasticity. Proc Natl Acad Sci 105(2):803-808

20. Goodstein DM, Shu S, Howson R, Neupane R, Hayes RD, Fazo J, Mitros T, Dirks W, Hellsten U, Putnam N, Rokhsar DS (2012) Phytozome: a comparative platform for green plant genomics. Nucleic Acids Res 40:D1178-D1186
21. Gutjahr C, Casieri L, Paszkowski U (2009) Glomus intraradices induces changes in root system architecture of rice independently of common symbiosis signaling. New Phytol 182:829-837
22. Haefele SM, Nelson A, Hijmans RJ (2014) Soil quality and constraints in global rice production. Geoderma 235–236:250-259
23. Hasegawa S, Thangaraj M, O'Toole JC (1985) Root behavior: field and laboratory studies for rice and non-rice crops. In: Soil Physics and Rice. International Rice Research Institute, Los Baños, Philippines. pp 383-395
24. Henry A, Cal AJ, Batoto TC, Torres RO, Serraj R (2012) Root attributes affecting water uptake of rice (Oryza sativa) under drought. J Exp Bot 63:4751-4763
25. Hill K, Porco S, Lobet G, Zappala S, Mooney S, Draye X, Bennett MJ (2013) Root systems biology: integrative modeling across scales, from gene regulatory networks to the rhizoshere. Plant Physiol 163:1487-1503
26. Huang CF, Yamaji N, Nishimura M, Tajima S, Ma JF (2009) A rice mutant sensitive to Al toxicity is defective in the specification of root outer cell layers. Plant Cell Physiol 50:976-985
27. Inthavong T, Fukai S, Tsubo M (2011) Spatial variations in water availability, soil fertility and grain yield in rainfed lowland rice: a case study from savannakhet province, Lao PDR. Plant Prod Sci 14:184-195
28. Iyer-Pascuzzi AS, Symonova O, Mileyko Y, Hao Y, Belcher H, Harer J, Weitz JS, Benfey PN (2010) Imaging and analysis platform for automatic phenotyping and trait ranking of plant root systems. Plant Physiol 152:1148-1157
29. Jannink JL, Lorenz AJ, Iwata H (2010) Genomic selection in plant breeding: From theory to practice. Brief Funct Genomics 9:166-177
30. Kano M, Inukai Y, Kitano H, Yamauchi (2011) Root plasticity as the key root trait for adaptation to various intensities of drought stress in rice. Plant Soil 342:189-196

31. Khalfaoui JL, Havard M (1993) Screening peanut cultivars in the field for root growth: a test by herbicide injection in the soil. Field Crops Res 32:173-179
32. Kervella J, Robert N, Fouilloux G (1993) Influence de la recombinaison sur la variabilité génétique. II. Etude par simulation. Agronomie 13:371-379
33. Khowaja FS, Norton GJ, Courtois B, Price AH (2009) Improved resolution in the position of drought-related QTLs in a single mapping population of rice by meta-analysis. BMC Genomics 10:e276
34. Kumar A, Dixit S, Henry A (2013) Marker-assisted introgression of major QTLs for grain yield under drought in rice. In: Varshney R, Tuberosa R (eds) Genomics Applications in Plant Breeding, Wiley-Blackwell Publishers, USA.
35. Lopes MS, Reynolds MP (2010) Partitioning of assimilates to deeper roots is associated with cooler canopies and increased yield under drought in wheat. Funct Plant Biol 37(2):147-156
36. Lorenz AJ, Chao S, Asoro FG, Heffner EL, Hayashi T, Iwata H, Smith KP, Sorrells ME, Jannink JL (2011) Genomic selection in plant breeding: knowledge and prospects. Adv in Agron 110:78-109
37. Lynch JP, Chimungu JG, Brown KM (2014) Root anatomical phenes associated with water acquisition from drying soil: targets for crop improvement. J Exp Bot Apr. Epub ahead of print
38. Lynch JP, Nielsen KL, Davis RD, Jablokow AG (1997) *SimRoot*: modelling and visualization of root systems. Plant Soil 188(1):139-151
39. Meijon M, Satbhai SB, Tsuchimatsu T, Busch W (2014) Genome-wide association study using cellular traits identifies a new regulator of root development in Arabidopsis. Nat Genet 46:77-81PubMed Abstract | Publisher Full Text
40. Meuwissen THE, Hayes BJ, Goddard ME (2001) Prediction of total genetic value using genome-wide dense marker maps. Genetics 157:1819-1829
41. Miao J, Guo D, Zhang J, Huang Q, Qin G, Zhang X, Wan J, Gu H, Qu LJ (2013) Targeted mutagenesis in rice using CRISPR-Cas system. Cell Res 23:1233-1236

42. Motte H, Vercauteren A, Depuydt S, Landschoot S, Geelen D, Werbrouck S, Goormachtig S, Vuylsteke M, Vereecke D (2014) Combining linkage and association mapping identifies RECEPTOR-LIKE PROTEIN KINASE1 as an essential Arabidopsis shoot regeneration gene. Proc Natl Acad Sci 111(22):8305-8310
43. Nicou R, Séguy L, Haddad G (1970) Comparaison de l'enracinement de quatre variétés de riz pluvial en présence ou absence de travail de sol. Agron Trop 25:639-695
44. Obara M, Wataru Tamura W, Ebitani T, Yano M, Sato T, Yamaya T (2010) Fine-mapping of qRL6.1, a major QTL for root length of rice seedlings grown under a wide range of NH4+ concentrations in hydroponic conditions. Theor Appl Genet 121:535-547
45. Orman-Ligeza B, Parizot B, Gantet P, Beeckman T, Bennett MJ, Draye X (2013) Post-embryonic root organogenesis in cereals: branching out from model plants. Trends Plant Sci 18:459-467
46. Ostlund G, Schmitt T, Forslund K, Kostler T, Messina DN, Roopra S, Frings O, Sonnhammer EL (2010) InParanoid 7: new algorithms and tools for eukaryotic orthology analysis. Nucleic Acids Res 38:D196-D203
47. O'Toole JC, Chang TT (1978) Drought and rice improvement in perspective. IRRI Res Pap Ser 14:27. Los Baños, Philippines
48. Pegoraro C, Farias DR, Mertz LM, dos Santos RS, da Maia LC, Rombladi CV, Costa de Oliveira A (2013) Ethylene response factors gene regulation and expression profiles under different stresses in rice. Theor Appl Genet 25:261-274
49. Pennisi E (2010) Sowing the seeds for the ideal crop. Science 327(5967):802-803
50. Postma JA, Dathe A, Lynch JP (2014) The optimal lateral root branching density for maize depends on nitrogen and phosphorus availability. Plant Physiol. [Epub ahead of print]
51. Pound MP, French AP, Atkinson JA, Wells DM, Bennett MJ, Pridmore T (2013) RootNav: navigating images of complex root architectures. Plant Physiol 162:1802-1814
52. Pradal C, Dufour-Kowalski S, Boudon F, Fournier C, Godin C (2008) OpenAlea: a visual programming and component-based software platform for plant modeling. Funct Plant Biol 35:751-760

53. Ranathunge K, Steudle E, Lafitte R (2003) Control of water uptake by rice (Oryza sativa L.): role of the outer part of the root. Planta 217:193-205
54. Ranathunge K, Kotula L, Steudle E, Lafitte R (2004) Water permeability and reflection coefficient of the outer part of young rice roots are differently affected by closure of water channels (aquaporins) or blockage of apoplastic pores. J Exp Bot 55:433-447
55. Rebouillat J, Diévart A, Verdeil JL, Escoute J, Giese G, Breitler JC, Gantet P, Espeout S, Guiderdoni E, Périn C (2009) Molecular genetics of rice root development. Rice 2(1):15-34
56. Rouard M, Guignon V, Aluome C, Laporte MA, Droc G, Walde C, Zmasek CM, Perin C, Conte MG (2011) Green) PhylDB v2.0: comparative and functional genomics in plants. Nucleic Acids Res 39:D1095-D1102
57. Schreiber F, Sonnhammer EL (2013) Hieranoid: hierarchical orthology inference. J Mol Biol 425:2072-2081
58. Shen L, Courtois B, McNally KL, Robin S, Li Z (2001) Evaluation of near-isogenic lines of rice introgressed with QTLs for root depth through marker-aided selection. Theor Appl Genet 103:75-83
59. Smith SE, Read DJ (2008) Mycorrhizal Symbiosis. Academic Press, New York.
60. Steele KA, Price AH, Witcombe JR, Shrestha R, Singh BN, Gibbons JM, Virk DS (2013) QTLs associated with root traits increase yield in upland rice when transferred through marker-assisted selection. Theor Appl Genet 126:101-108
61. Steele KA, Virk DS, Kumar R, Prasad S, Witcombe JR (2007) Field evaluation of upland rice line selected for QTLs controlling root traits. Field Crop Res 101:180-186
62. Sukumar P, Legué V, Vayssières A, Martin F, Tuskan GA, Kalluri UC (2013) Involvement of auxin pathways in modulating root architecture during beneficial plant–microorganism interactions. Plant Cell Environ 36:909-919
63. Swamy BPM, Ahmed HU, Henry A, Mauleon R, Dixit S, Vikram P, Ram T, Verulkar SB, Perraju P, Mandal NP, Variar M, Robin S, Chandrababu R, Singh ON, Dwivedi JL, Das SP, Mishra KK, Yadaw RB, Aditya TL, Karmakar B, Satoh K, Moumeni A, Kikuchi

S, Leung H, Kumar A (2013) Genetic, physiological, and gene expression analyses reveal that multiple QTL enhance yield of rice mega-variety IR64 under drought. PLoS One 8:e62795. doi:10.1371/journal.pone.0062795

64. Topp CN, Iyer-Pascuzzi AS, Anderson JT, Lee CR, Zurek PR, Symonova O, Zheng Y, Bucksch A, Mileyko Y, Galkovskyi T, Moore BT, Harer J, Edelsbrunner H, Mitchell-Olds T, Weitz JS, Benfey PN (2013) 3D phenotyping and quantitative trait locus mapping identify core regions of the rice genome controlling root architecture. Proc Natl Acad Sci 110:E1695-E1704
65. Trachsel S, Kaeppler SM, Brown KM, Lynch JP (2011) Shovelomics: high throughput phenotyping of maize (Zea mays L.) root architecture in the field. Plant Soil 341:75-87
66. Tracy SR, Roberts JA, Black CR, McNeill A, Davidson R, Mooney SJ (2010) The X-factor: visualizing undisturbed root architecture in soils using X-ray computed tomography. J Exp Bot 61:311-313
67. Uga Y, Sugimoto K, Ogawa S, Rane J, Ishitani M, Hara N, Kitomi Y, Inukai Y, Ono K, Kanno N, Inoue H, Takehisa H, Motoyama R, Nagamura Y, Wu J, Matsumoto T, Takai T, Okuno K, Yano M (2013) Control of root system architecture by *DEEPER ROOTING 1* increases rice yield under drought conditions. Nat Genet 45:1097-1102
68. Wang LK, Chu HW, Li ZY, Wang J, Li JT, Qiao Y, Fu YR, Mou TM, Chen CL, Xu J (2014) Origin and development of the root cap in rice. Plant Physiol 114.240929. Epub ahead of print
69. Wei FJ, Droc G, Guiderdoni E, Hsing YI (2013) International consortium of rice mutagenesis: resources and beyond. Rice (N Y) 6:39
70. Woll K, Borsuk LA, Stransky H, Nettleton D, Schnable PS, Hochholdinger F (2005) Isolation, characterization, and pericycle-specific transcriptome analyses of the novel maize lateral and seminal root initiation mutant rum1. Plant Physiol 139:1255-1267
71. Yambao EB, Ingram KT, Real JG (1992) Root xylem influence on the water relations and drought resistance of rice. J Exp Bot 43:925-932
72. York LM, Nord EA, Lynch JP (2013) Integration of root phenes for soil resource acquisition. Front Plant Sci 4:355

Chapter 8

Planted Forest Development in Australia and New Zealand: Comparative Trends and Future Opportunities

David Rhodes[1] and Michael Stephens[2]

[1]New Zealand Forest Owners Association, PO Box 10986, Wellington 6143, New Zealand

[2]Australian Forest Products Association, PO Box 239, Deakin West ACT 2600, Australia

ABSTRACT

Background

The areas of commercial wood plantations in Australia and New Zealand are 2.01 and 1.71 million hectares respectively, representing the predominant type of planted forest.

Methods

This paper reviews the status of planted forests in Australia and New Zealand and discusses opportunities for further development taking into account economic, social and environmental factors. These factors include regional wood demand and supply conditions, emerging markets for ecosystem services, climate change impacts and social attitudes toward planted forests.

Results

Both countries have followed a similar pattern of state intervention for the development of wood plantations, relying on government ownership and direct incentives such as taxation concessions that have gradually shifted to the use of more indirect policies. The main drivers for wood plantations were as an additional source of wood outside the natural forest, improved productivity and regional development. More recent policy has focused on the privatisation of forest assets and the promotion of ecosystem services such as carbon sequestration. To date, plantings for ecosystem services such as carbon sequestration has been limited.

Conclusions

It is argued that better design of policy settings is needed to capture the carbon abatement benefits and other ecosystem services from planted forests, which may help overcome current investment hurdles. Careful attention will also need to be applied to managing future climate change risks as well as to realising the opportunities from forecast growth in wood and fibre product demand particularly in the Asian region.

INTRODUCTION

The Food and Agriculture Organisation (FAO) (2010) defines a planted forest as those forests 'composed of trees established through planting and/or through deliberate seeding of native or introduced species', which includes forest plantations[a] and planted semi-natural[b] forests. Globally, planted forests have increased in area by just over 8% per cent between 2005 and 2010 to a total of 264 Mha (FAO 2010).

[a] Forest plantations, a subset of all planted forests, are defined as forests of introduced species and in some cases native species, established through planting or seeding, with few species, even spacing and/or even-aged stands (FAO 2006a).

[b] Semi-natural forests are defined as forests of native species, established through planting, seeding or assisted natural regeneration. This definition includes areas under intensive management where native species are used and deliberate efforts are made to increase/optimise the proportion of desirable species, thus leading to changes in the structure and composition of the forest, with possible presence of naturally regenerated trees from other species than those planted/seeded. This may include areas with naturally regenerated trees of introduced species and areas under intensive management where deliberate efforts, such as thinning or fertilising, are made to improve or optimise desirable functions of the forest (FAO 2006b).

Within Oceania, there is approximately 4 Mha of planted forest, with Australia and New Zealand comprising over 92% of this area (Carle and Holgrem 2008). Other countries in the region that have developed a planted forest resource include Fiji and Papua New Guinea. In this paper, planted forest development is discussed with respect to Australia and New Zealand, as the major plantation forested countries in Oceania.

This paper reviews the status of planted forests in Australia and New Zealand and provides a comparative analysis of recent trends and opportunities for their development, taking into account economic, social and environmental factors. These factors include regional wood demand and supply conditions, emerging markets for ecosystem services such as carbon sequestration, climate change impacts and social attitudes toward planted forests. These factors represent a number of challenges and opportunities for planted forest development which to date has focused primarily on wood production. It is argued that better design of policy settings is needed to capture the multiple carbon abatement benefits and other ecosystem services from planted forests, which may help overcome current investment hurdles for new wood plantations. The proximity of Australia and New Zealand to Asia is also discussed in terms of forecast growth in regional demand for forest products and implications for plantation development and associated wood processing.

WOOD PLANTATIONS

Planted forests currently play an important role in Australia and New Zealand as a source of wood and fibre supply, partly in response to the declining availability of industrial wood from natural forests. The plantation estates of New Zealand and Australia produce roughly the same volume of logs per annum at 26 million and 27 million cubic metres respectively. In terms of their relative contribution to total annual log supply, wood plantations contribute 99.9% in New Zealand and over 80% in Australia (Australian Bureau of Agricultural and Resource Economics and Sciences (ABARES) 2014; New Zealand Forest Owners Association 2013). This reflects a contraction over time in natural forest availability for wood production and a maturing plantation resource for harvest.

In Australia, the area of commercial wood plantation is just over 2 Mha, with 1.02 Mha of coniferous plantation and 980 000 ha of broadleaved plantation (Gavran et al. 2012). Australia's commercial plantations are located mainly in the north-eastern coastal, south-western, southern and eastern regions of Australia.

The coniferous plantation is dominated by *Pinus radiata* D. Don or radiata pine (75%) and *Pinus elliottii* Engelm. and *Pinus caribaea* Morlet hybrids known as southern pines (15%). Most radiata pine is located in Victoria, New South Wales and Tasmania, and most southern pines are located in south east Queensland. *Pinus pinaster* Aiton, or maritime pine, is the dominant coniferous plantation species in Western Australia. The broadleaved plantation is dominated by *Eucalyptus globulus* Labill. (55%) and *Eucalyptus nitens* H. Deane & Maiden (24%), which are primarily managed for short-rotation pulpwood production.

In the seasonally dry tropical regions of northern Australia, there is also around 10 000 ha of planted African mahogany (*Khaya senegalensis* (Desr.) A.Juss.) which has considerable potential as a high value timber plantation species (Dickinson et al. 2011), as well as over 7 600 ha of Indian sandalwood (*Santalum album* L.) which is used in the global fragrance market.

In New Zealand, there is 1.71 Mha of wood plantation which is dominated by exotic species representing 99% of total planted area, including radiata pine (90%) and *Pseudotsuga menziesii* or Douglas-fir (6%). The remainder of the plantation estate is comprised of a mixture of Cypress species and other exotic softwoods and hardwoods (New Zealand Forest Owners Association 2013).

HISTORICAL DEVELOPMENT

Planted forest development in Australia and New Zealand has followed a similar pattern in terms of direct state intervention for the development of wood plantations over the last 150 years or so, which has gradually shifted to more indirect policies for the improved operation of markets and the promotion of ecosystem services such as carbon sequestration. These policies have resulted in a sizeable wood plantation resource and processing industry, whereby the forest products industry contributes $NZ4.5 billion and $A6.3 billion to national gross domestic product (GDP) in New Zealand and Australia respectively. The primary focus on wood production reflects international trends for planted forests, where it is estimated that 76% of global planted forests have production as their primary goal (FAO 2010).

Stephens and Grist (2014) and Ferguson (2014a) describe early state intervention in Australia for wood plantations via direct government ownership that has shifted to indirect policies such as the Plantations 2020 Vision. The Plantations for Australia: 2020 Vision, launched by the federal and state governments in 1997, aimed to enhance wealth creation and international competitiveness through a notional target of trebling the area of wood plantations to 3 Mha by 2020. This policy has focused on improving competitiveness through research and development, improving information on markets and prices and streamlining planning and regulation.

Large-scale wood plantation development began in the 1960s, through the use of concessional loans from the federal government to the state governments which established and managed the development of long-rotation (e.g. 35 year) coniferous plantation. The major goals of this expansion were to promote regional development and increase wood productivity and self-sufficiency, given the adverse terms of trade in forest products at that time (Ferguson 2014a). These loans led to the establishment of 1 Mha of coniferous plantation for integrated sawlog and pulpwood, used for domestic sawn timber and pulp and paper production (Figure 1). The other major trend that occurred from the 1990s was the emergence of Managed Investment Schemes (MIS) which took advantage of existing taxation incentive provisions and pooled capital from a large number of small investors. This led to the rapid rise of mostly short rotation hardwood (*Eucalyptus* spp.) plantations for woodchip export markets, which increased from a relatively low base to over 800 000 ha by 2012.

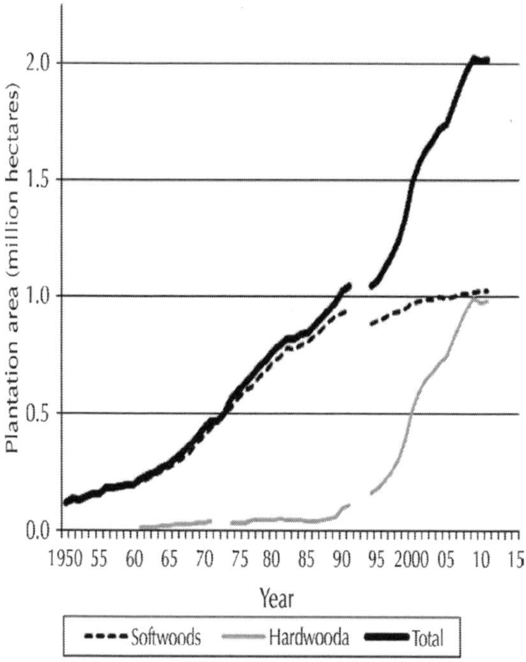

Figure 1: Plantation forest area in Australia, 1950-2011 (*Source*: ABARES.).

However, problems with the structure of the financial arrangements and issues with access to suitable land led to the collapse of many MIS companies, with some areas planted on uneconomic sites. Following the onset of the Global Financial Crisis in 2007-8 and the collapse of many MIS companies, most of this estate has been acquired by institutional investors. Further rationalisation of the former MIS estate is projected, with some areas not expected to be commercially replanted as a consequence of being established on poor quality sites or other factors such as long distances to markets. Some companies estimate that up to 30% of the former MIS hardwood plantation estate may not be replanted (Ferguson 2014b), with this land being converted back to agriculture following harvest (Stephens and Grist 2014).

Other forests have been planted for ecosystem services such as soil and water conservation, biodiversity and carbon sequestration. There is estimated to be 153 000 ha of non-industrial planted forest in Australia, some of which is used for environmental purposes and farm forestry activities (Montreal Process Implementation Group for Australia and National Forest Inventory Steering Committee 2013). Consultants

(URS Forestry 2008) reported 155 000 ha of farm forestry plantings in Australia based on a 2008 survey, which included some large-scale plantings as part of leasehold arrangements with MIS companies. In addition to wood and agricultural productivity benefits (e.g. shade, shelter), a proportion of these plantings is for ecosystem services such as biodiversity and soil conservation, consistent with earlier surveys of farmers undertaking tree planting activities (Wilson et al. 1995).

Similarly, in New Zealand, there was a period of direct state intervention for wood plantation promotion from the 1960s through to the early 1980s. Over this period the plantation forest estate expanded by almost 700 000 ha through a range of programmes, including forestry encouragement grants, low interest loans and taxation deductions of up to 100% of planting costs to support private investment, as well as government planting programmes through the New Zealand Forest Service (Rhodes and Novis 2002, de Fegeley et al. 2011).

Other priorities for planted forests in New Zealand have included land rehabilitation and control of soil and water erosion from previous vegetation clearing for agriculture, particularly on steep slopes. In the Gisborne district of the North Island, for example, there has been a history of planting to combat erosion through Government-owned plantations and grants to private landowners. This has included the East Coast Forestry Project, which has assisted landowners to plant trees since 1992 for multiple goals including erosion control, wood production and employment (Rhodes 2001). Reforestation has been used to improve water quality that declined due to eutrophication from intensive agriculture (Maclaren 1996), and more recently, for the development of a carbon trading scheme.

Ownership Trends

A significant feature of the wood plantation estate in Australia has been the shift in ownership from the public sector (i.e. state governments) to the private sector (Figure 2). The harvest and management rights to most of the coniferous plantations were sold in the 2000s to private institutional investors as part of the commercialisation of government owned assets. The management of these plantations is undertaken by timber investment management organisations (TIMOs) on behalf of investors, consistent with recent trends in North America. The other trend has been the acquisition by TIMOs of the hardwood eucalypt plantations formerly owned or managed by the failed MIS companies.

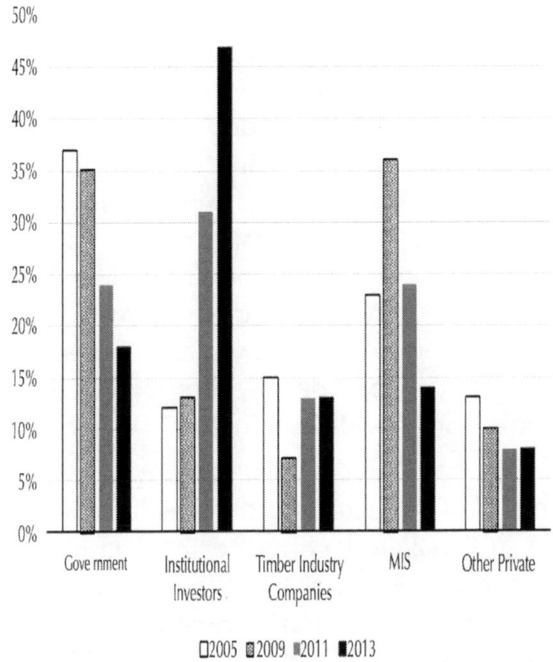

Figure 2: Plantation ownership in Australia (*Source*: ABARES; Australian Forest Products Association).

In New Zealand, there is 1.71 Mha of wood plantation. Fifteen companies manage plantation forest in excess of 20 000 ha with the largest being just over 235 000 ha (New Zealand Forest Owners Association 2013). These fifteen companies manage 62% of the plantation estate. Six companies identify themselves as TIMOs, managing 389 000 ha, and 13 companies identify as property management companies managing just over 1 Mha. Maori and the Crown own considerably more land than forest, with Maori leasing 326 000 ha and the Crown leasing 219 000 ha to forest management companies. Another feature of New Zealand's ownership pattern is that 629 000 ha (or just over 37%) is owned by private individuals and landowners with less than 10 000 ha. An increasing proportion of this area will reach harvest age in the next 5 to 10 years.

REGIONAL WOOD DEMAND AND SUPPLY

Given their close proximity and on-going forest products trade in Asia, wood demand and supply in the region has a direct influence on profitability and investment in wood plantations in Australia and New Zealand. There is also considerable intra-regional trade between the two countries. Australia is a net importer in forest products of around $2 billion per annum, comprising mostly imports of paper and paperboard from China and sawn timber from New Zealand. Australia has also been a large exporter of hardwood woodchips to Japan and China for pulp and paper production, which has averaged over $A850 million per annum over the past decade. Log exports to China have also increased markedly from 201 000 m^3 in 2002-03 to over 1.2 million m^3 in 2012-13 (ABARES 2014).

New Zealand is a large exporter of forest products, worth around $NZ 4.5 billion in 2012, reflecting the importance of the sector to the national economy at 3.3% of GDP (New Zealand Forest Owners Association 2013). In 2012, the major forest product export partners by value included China ($1.45 billion), Australia ($775 m), Japan ($542 m) and Korea ($392 m).

In terms of demand for forest products, China's influence is increasingly significant for Australia and New Zealand. Over the past few decades, woodchip exports to Japan have been the major market for Australian hardwood plantation pulpwood. However, China's share of woodchip exports value from Australia has grown from less than 1% in 2001-02 to around 20% in 2011-12 (ABARES 2014). There has also been a fundamental shift in overseas market share for New Zealand's forest products, particularly the rise of China to 67% of log and pole exports in 2012. In 2012, China represented 34% of New Zealand's total forest product exports by value, compared to just 4% in 1992 when Australia's share was 34% of forest product exports.

The demand outlook in Asia is influenced by economic growth assumptions and underlying social and environmental policies. Buongiorno et al. (2012) and Indufor (2012) present a range of global demand scenarios in terms of population growth, GDP and adoption of environmental and climate change policies. A consistent trend in these scenarios is the high economic growth in China and India over the next few decades and their forecast position as net forest product importers. TheAsia Pacific Forestry Commission (2010) forecast demand for industrial round wood to increase from 317 million m³ in 2005 to 550 million m³ in 2020, with East Asia, especially China, accounting for most of this consumption. Similar analysis has shown that growth in construction is expected to be strong in China, the United States and India, with a 70% increase in construction output to $15 trillion worldwide by 2025, including the need for 270 million new homes in China and India by 2015 (Global Construction Perspectives and Oxford Economics 2013).

Global wood pellet production, as a source of fuel for energy, has grown from just less than 2 Mt in 2000 to around 16 Mt in 2010 (Lamers et al. 2012). Global wood pellet demand is expected to grow strongly to around 40-50 Mt per annum by 2020 with Europe accounting for a large share of consumption. The Asian region, which is a much closer market than Europe for Australia and New Zealand, is expected to rise from less than 1 Mt in 2010 to around 10-15 Mt by 2020 (International Energy Agency 2011).

The outlook for Australian and New Zealand wood plantations is also affected by local demand, which is largely influenced by Australian economic activity as a medium sized economy of almost 23 million people (Australian Bureau of Statistics 2014). Recent forecasts of domestic consumption suggest a moderate increase for most forest product categories (Table 1), based on assumed economic growth, rates of house

Table 1: Long-term consumption forecasts: Australia

	Time period			
Product	**2010-11** a	**2012-13** a	**2029-30**	**2049-50**
Sawn wood (Mm^3)	5.0	5.1	6.0	6.5
Wood based panels (Mm^3)	2.0	1.7	3.1	4.3
Paper and paperboard (Mt)	4.0	3.6	5.8	7.1

Source: ABARES (2013, 2014). a = actual consumption. Mm^3 = million cubic metres. Mt = million tonnes.

Rhodes and Stephens *New Zealand Journal of Forestry Science* 2014 44(Suppl 1):S10z, doi:10.1186/1179-5395-44-S1-S10

The consumption of forest products in New Zealand has similarly remained steady over the past few years, albeit from a smaller population base than Australia (Table 2). As a net exporter of forest products with a relatively small domestic market, growth opportunities primarily lie in the global market.

Table 2: Consumption of forest products: New Zealand

	Year		
Product	**2006**	**2010**	**2014**
Sawn wood (Mm^3)	2.5	1.9	2.2
Wood based panels (Mm^3)	0.7	0.4	0.8
Paper and paperboard (Mt)	0.7	0.9	0.8

Source: Ministry for Primary Industries (2014).

Rhodes and Stephens New Zealand Journal of Forestry Science 2014 44(Suppl 1):S10, doi:10.1186/1179-5395-44-S1-S10

On the supply side, Carle and Holgrem (2008) undertook a global survey of planted forests in 61 countries and predicted around a one third increase in total area over the next two decades, to reach 345 Mha by 2030 (Table 3). By comparison, Oceania was predicted to have an increase in area of around 11% or 400 000 ha by 2030.

Table 3: World and Oceania planted forest area, 2005 (actual) and 2030 (forecast)

Year	Region	Softwoods				Hardwoods						Total
		***Pinus* spp.**		**Other spp.**		***Acacia* spp.**		***Eucalyptus* spp.**		**Other spp.**		
2005	Oceania	**Mha**	**%**	**Mha**	**%**	**Mha**	**%**	**Mha**	**%**	**Mha**	**%**	**Mha**
		2.7	68	0.2	5	0.0	0	0.5	13	0.2	5	3.6
	World	73	28	64	25	9	0.3	14	5	100	38	261
2030	Oceania	2.9	69	0.2	5	0.0	0	0.8	19	0.3	7	4.2
	World	94.9	28	83.0	24	9.9	0.3	21.4	6	135.5	39	344.6

Source: Carle & Holgrem (2008). Note: Mha = millions of hectares.

Rhodes and Stephens New Zealand Journal of Forestry Science 2014 44(Suppl 1):S10, doi:10.1186/1179-5395-44-S1-S10

According to Indufor (2012), the total area of industrial forest plantation in the world was 54.3 Mha in 2012 and may increase to 91 Mha by 2050. Indufor (2012) estimated that the area of industrial plantation in Asia and Latin America could double by 2050 with some expansion in Africa and Oceania.

In Australia, a recent survey of plantations and log availability has estimated that the plantation area may remain relatively constant at 2 Mha up to 2019-20 (Gavran et al. 2012). However, log volumes are expected to increase as a result of the maturing resource to average over 29 million m^3 per year in 2015-19 and reach an average annual peak of 33 million m^3 in 2030-34, largely as a result of an increase in hardwood pulpwood with softwood volumes to remain relatively static (Figure 3).

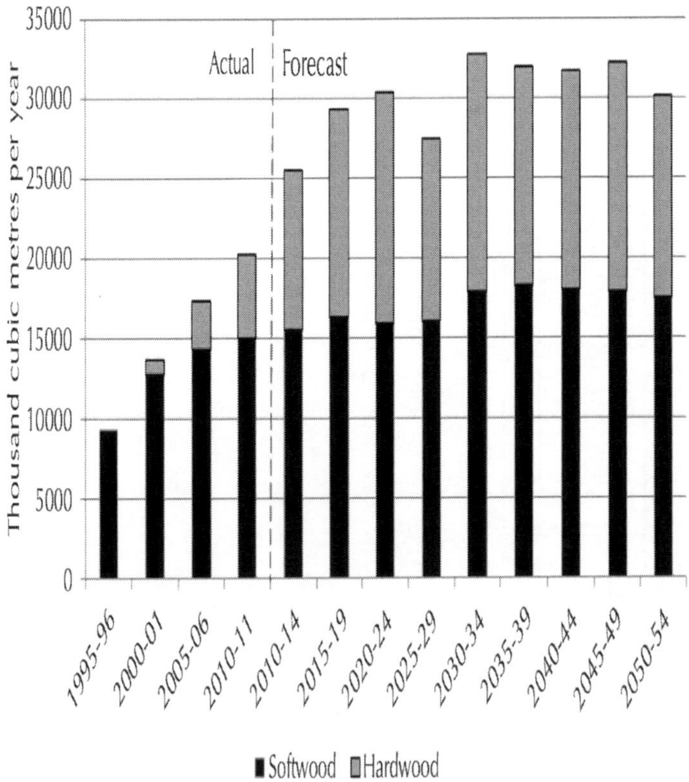

Figure 3: Australian plantation log availability 1990-2050 (*Source:*Gavran et al. (2012); ABARES (2014)).

While providing a useful basis for national wood assessments, Ferguson (2014b) has identified some limitations in plantation forecasting in Australia, including the need for more robust sampling for growth and yield and information on replanting assumptions. This is particularly relevant in the case of the rationalisation of the former MIS plantations, with some areas not to be replanted for the reasons identified above.

In New Zealand, the harvest level is around 30 million m^3 of wood from plantations (Ministry for Primary Industries, 2014). Although market conditions eased in 2014, the steady increase in harvest volumes over the past five years is forecast to continue in the coming years to reach around 35 million m^3 by 2022 (Wood Council of New Zealand 2012). Most of this increase is to come from small growers who established forests in the early 1990s and actual harvest levels will depend upon market conditions and the decisions of small-scale owners (Ministry for Primary Industries 2013a).

ECOSYSTEM SERVICES

The predominant purpose of global planted forests has been for wood and fibre production (FAO 2010), which has also been the case in Australia and New Zealand. However, it is expected that the relative importance of planted forests for environmental purposes will increase in the future, given their potential to contribute to a range of ecosystem services such as rehabilitation of degraded sites, protection of soil and water, regulation of watersheds, biodiversity and carbon sequestration (Bahaus et al. 2012). This has been the case in China, for example, which has tripled the area of planted forest between 1990 and 2010 to 60 Mha for the large scale control of desertification (FAO 2010).

Planted forests can have both positive and negative environmental consequences such as the introduction of pests and water-use impacts, but through careful design and planning can help address landscape management issues (Nambiar and Ferguson 2005). When assessing the scope for forest plantings in Australia and New Zealand for ecosystem services, it is important to note the previous environmental degradation of land, water and vegetation, primarily as a result of agricultural development and other settlement patterns since the arrival of Europeans in the late 18th century. In Australia, it is estimated that around 13% or

100 Mha of land has been cleared of native vegetation, mostly eucalypt woodlands, with a further 65% subject to disturbance pressures such as livestock grazing (State of the Environment Committee 2011). This has led to a range of natural-resource management problems, including soil and water erosion, sedimentation, acidification, dry-land salinity and loss of plant and animal diversity.

Plantings for ecosystem services in Australia have evolved from government sponsored schemes (e.g. grants, community education) in the 1980s for land rehabilitation and biodiversity to broader climate change mitigation schemes. Stephens and Grist (2014) identified three distinct trends for ecosystem services plantings, comprising:

- programmes for landscape restoration and farm productivity (1980s);
- greater recognition of biodiversity values and landscape connectivity from revegetation and planted forests (1990s); and
- the emergence of carbon markets for forestry and land sector activities (2000s).

In both Australia and New Zealand, research has found that plantations can provide habitat benefits for a number of species compared to agricultural and pastoral land (Maclaren 1996, Lindenmayer and Hobbs 2004, Kavanaugh et al. 2007, Hock et al. 2009, Baral et al. 2013). There has also been considerable interest in the development of market-based mechanisms to capture ecosystem services such as biodiversity for forest owners. Keenan et al. (2004) identified the early potential for market based mechanisms to encourage small scale private landowners in Australia to deliver environmental services, such as to retain existing forest, establish forests on previously cleared land and integrate trees more fully with farming systems. It was argued that while forest owners can make a profit from the sale of wood and other commercial products, environmental services were not effectively valued or traded as market goods. However, it has been difficult for plantation owners and investors to generally capture the non-market benefits from ecosystem services, given the lack of market development and range of site-specific outcomes and conditions.

In a dry continent such as Australia, it is recognised that plantations, like all forms of agricultural land use, intercept water and that plantations generally use more water than pasture (Zhang et al. 2003). Plantations can also play an important role in improving water quality through reduced use of chemicals and pesticides, reduced erosion and soil salinity and flood control (Zhang et al. 2007). Water policy in Australia has therefore been directed at achieving an appropriate balance between managing the impacts of plantations on both water yield and quality. The interaction of plantations on both these variables is important, as the costs and benefits of reduced run-off and/or lower water tables will depend on the salinity of the groundwater, its potential to contribute to waterlogging or inundation and its potential for productive use (Zhang et al. 2007). In the south-east region of South Australia, for example, groundwater is a highly valued freshwater resource, while in Western Australia there is high saline groundwater and dry-land salinity issues from previous land clearing. At a national level, dry-land salinity is a particularly important water management issue, which arises from the removal of vegetation and rising saline water tables that causes land to be unproductive. The National Land and Water Resources Audit (2001) estimates that 5.7 Mha of land is adversely affected by dry-land salinity which is forecast to increase to 17 Mha by 2050 in the absence of amelioration.

Zhang et al. (2007) suggest that afforestation could play a significant role to reduce dry-land salinity, given the ability of trees to manage recharge through their deep rooting habit and high perennial water use. Considerable work has been undertaken in assessing and trialling salt tolerant tree species to ameliorate dry-land salinity in a range of geographic regions and sites, which can also produce other related benefits such as the harvest of biomass for fodder or bioenergy (Dale and Dieters 2007, Sochaki 2012) and deliver multiple wood, water quality and carbon sequestration benefits (Townsend et al. 2012).

Stephens and Grist (2014) reviewed the market potential for delivering joint wood production and ecosystem services from planted forests in Australia, including biodiversity, water quality and carbon sequestration. They concluded that there was limited scope, at least in the short term, for biodiversity payments given the complexity of capturing the benefits and transaction costs. By comparison, dry-land salinity mitigation was considered important in a regional context where markets do exist at a catchment scale and there is greater market development as part of the national water market for consumptive uses. Carbon markets for sequestration were considered most prospective, given the development of a domestic crediting and verification mechanism and the fungibility of tradable carbon credits.

In New Zealand, Hock et al. (2009) assessed the potential for ecosystem services from planted forests and reached similar conclusions regarding a lack of information and tradable markets for services other than for carbon sequestration. Despite commercial limitations to capturing ecosystem services, Yao et al. (2014) conducted a choice-modelling experiment and found that biodiversity enhancement in large-scale planted forests is valued in the community, and could potentially form part of a suite of future goods and services.

Carbon Markets

Planted forests can provide multiple carbon sequestration and fossil fuel displacement benefits, through the carbon stored in forests and harvested wood products, the substitution of more energy-intensive products with wood products and the use of woody biomass for renewable energy (Malsheimer et al. 2011, Lippke et al. 2011, Oliver et al. 2014).

Both New Zealand and Australia have been early movers in terms of the implementation of domestic carbon market based schemes with mixed results in terms of promoting new plantings. In Australia, this has included the New South Wales Greenhouse Gas Reduction Scheme (GGAS) and the Carbon Farming Initiative, the latter which links as a land crediting mechanism to the national Emissions Reduction Fund (ERF). The ERF is a reverse auction market mechanism to buy low-cost emission reduction activities in the Australian economy, with the Australian Government initially allocating $1.55 billion to the Fund over a three year period commencing 1 July 2014 (Commonwealth of Australia 2013).

Mitchell et al. (2012) provided a summary of the evolution of carbon forestry schemes and their uptake in Australia, which has resulted in the establishment of 65 000 ha. The main schemes have generally been aimed at Kyoto Protocol compliant activities, such as afforestation and reforestation activities on cleared agricultural land post-1990. Most carbon plantings have been for not-for-harvest projects, either mixed plantings of eucalypt species, or plantings of mallee eucalypts in low rainfall areas that are multi-branched trees with a height rarely exceeding 6 metres. Overall, the uptake of wood plantations has been minimal given uncertainty in carbon market settings and scheme design issues (Mitchell et al. 2012). These issues have included the treatment of additionality, carbon pools (e.g. inclusion of wood products) and crediting periods (Stephens and Grist 2014). For example, the proposed design of the ERF is to have five-year contracts, which may disadvantage land based sequestration such as forestry projects, which typically have low early growth but higher longer term sequestration rates.

However, given the large area of previously cleared land, there has been strong interest in the potential for sequestration from planted forests under a range of carbon price and land availability scenarios. Several spatial analyses at a national scale have assessed the economic potential for plantations under a carbon market using physiological growth models and agricultural land opportunity costs, with estimates ranging from less than 350 000 ha up to 9 Mha by 2050 (Stephens and Grist 2014). In the absence of a carbon price, Paul et al. (2013) found that establishment of new plantations for wood production on agricultural land was not presently viable in Australia given current costs of establishment. However, on average, for scenarios of hardwoods (*Eucalyptus* spp.) and softwoods (e.g. *Pinus radiata*), a payment of about $A10 and $A30/t CO_2 respectively would make plantation expansion viable assuming a 4% carbon price increase per year, with an initial price above $A20/t CO_2 needed in the majority of scenarios for south-eastern Australia. The scenarios included short- and long-rotation plantations for pulpwood and sawlogs and the crediting of carbon stored in wood products and soils. The findings are consistent with de Fegeley et al. (2011) who suggested that without a carbon price new wood plantations would not be able to meet the investment hurdle rate.

In New Zealand, an emissions trading scheme was introduced in 2009 and forestry was the first sector to be included, generating an initial stimulus of forest plantings for sequestration. In 2011, for example, new forest plantings were almost double the previous year at 12 000 ha (New Zealand Forest Owners Association 2013). However, scheme design issues have since eroded the incentives for planted forests through low domestic carbon prices and high volatility and risk brought about by a series of amendments to the scheme. The depressed New Zealand carbon prices are attributed to the partial sectoral coverage of the scheme and strong linkage to uncapped international carbon trading of Kyoto compliant units, which have also fallen significantly in recent years. Under the New Zealand scheme, any credits are also balanced against any liability from the net release of carbon at harvest.

The ability of a carbon price to drive land use change is also dependent on the agricultural revenues foregone and the objectives of the owner (West et al. 2011). Manley and Maclaren (2009) estimate that a price of \$NZ20/t CO_2 could improve the internal rate of return on a typical radiata pine plantation from 5.4% to 11.0% and could stimulate up to 25 000 ha/annum of new planting. In addition, carbon prices can result in modified management practices in new and existing planted forests. Hale et al. (2014) modelled the trade-offs between timber harvest and carbon sequestration and found that while more productive regions of New Zealand such as the Central North Island and Northland have a greater capacity as a carbon sink, it is the less productive regions that have a comparative advantage in carbon sequestration in terms of a lower cost of wood production revenue forgone. Various treatments such as switching from pruning to plant-and-leave or other harvest constraints may be optimal for a range of assumed carbon prices.

Forest Management

The legal and institutional policy framework for sustainable forest management in Australia and New Zealand is well described in McDermott et al. (2010) and includes government environmental and planning regulation and codes of forest practice for plantation operations. These codes and regulations prescribe operational practices such as stream buffer areas that cannot be planted, how to build and maintain roads, use of fertilisers and pesticides, erosion control and fire protection requirements. In addition, there is high adoption of voluntary internationally recognised certification schemes for sustainable forest management. In Australia, around three-quarters of production forests are certified under either, or both, the Forest Stewardship Council (FSC) or the Programme for the Endorsement of Forest Certification (PEFC) via the Australian Forestry Standard. Similarly, New Zealand plantation forests began obtaining FSC certification during the mid-1990s and almost 60% of New Zealand's planted forests are certified under FSC.

CLIMATE CHANGE IMPACTS

Climate change can have an important impact on planted forest development. Several studies have looked at the possible impacts of climate change on planted forests in Australia and New Zealand (Watt et al. 2008, Battaglia et al. 2009, Booth et al. 2010, Singh et al. 2010, ABARES 2011), including the effects of predicted changes in CO_2 concentration, temperature and rainfall, as well as the threats posed by pests, diseases, weeds, fire and drought.

Climate forecasts vary, but generally predict that New Zealand could be 2°C warmer from 1990 to 2090 and Australia could be warmer by between 0.4 and 2°C by 2030 and between 1 and 6°C by 2070. Variable rainfall trends are also predicted, with the western side of New Zealand's two large islands predicted to have increasing rainfall while the eastern side less (Watt et al. 2008). In Australia, it is predicted that southern Australia will have less rainfall with some increases in northern Australia.

A key consideration is the impact of climate change on forest growth. Plantation growth may increase from rising levels of atmospheric carbon dioxide but may decrease from rising temperature and increased water loss (evapotranspiration). Increased temperature may also increase productivity in some regions and for some species, such as an extended growing season in higher altitude regions and reduced frost. Battaglia et al. (2009) modelled the effects of climate forecasts on plantation productivity in Australia and found a net increase in growth for some major growing regions and for other regions a net decrease. Drought-induced tree mortality was also identified as a significant risk in some drought-prone regions. Watt et al. (2008) identified a general gain in forest growth in New Zealand given relatively high water availability but increased drought risk in some areas.

The other major consideration is the impact of the secondary effects of climate change. In New Zealand, there is forecast to be stronger winds and more high intensity rainfall as well as less risk of fungal disease in drier areas but increased risk from fire and insect attack. Fire risk is also forecast to be significant in Australia. Hennessy et al. (2007) predicted that the frequency of very high and extreme fire danger days in Australia is likely to rise 4-25% by 2020 and 15-70% by 2050. Stephens (2010) has identified that these risks raise significant fire management issues at an individual stand and landscape scale, considering that planted forests occupy less than 2% of the total forest area in Australia of 125 Mha. Cyclones in northern Australia are also predicted to increase in frequency and impact on areas further south of current ranges. Higher susceptibility to pest and disease outbreaks is predicted, resulting from changes in both pathogen vectors and the health of trees influenced by climatic factors, although these factors are the least understood in both Australia and New Zealand (Watt et al. 2008, Singh et al. 2010).

Overall, Australian plantation forests are classified as moderately vulnerable to climate change given that the major commercial species are grown over wide climatic ranges (Booth et al. 2010). It is also recognised that there is large capacity for adaptation. The main management responses are described by Booth et al. (2010), including: spacing and thinning, watering and nutrient regimes, fire management, pest, disease and weed management, use of genotypes and site selection, climate monitoring and establishment practices. Adaptive forest management is regarded as an important tool for plantation managers and the industry has responded in Australia through the development of capacity building tools such as a plantation forest industry climate change adaptation handbook (Stephens et al. 2012).

SOCIAL ATTITUDES AND LAND ACCESS

Commercial afforestation of agricultural land is often associated with social conflict over the perceived environmental, economic and social impacts of plantations (Schirmer 2007). Planted forest development in Australia and New Zealand has attracted some community concerns, primarily regarding competition for agricultural land, employment impacts and environmental issues. Williams (2008) undertook a survey of community attitudes to wood plantations in Tasmania and the south-western regions of Western Australia and found that respondents tended to value agriculture land use higher than for plantation forestry. Many people believed plantations offered benefits to companies and individuals but only limited regional employment and economic benefits. Similarly, Langer and Barnard (2002) surveyed residents in the Gisborne East Coast region of New Zealand and identified that community concerns about forest expansion included lack of employment generation and the perceived loss of good farming land. These views have been exacerbated in Australia from the collapse of many MIS companies and concerns over the rapid expansion of plantations prior to their collapse and upward pressure on land prices in some areas (Ferguson 2014a). Other issues have included perceptions of naturalness and implications for planted forest management in New Zealand (Fairweather and Swaffield 2003).

Research has shown that plantation forestry can generate a range of employment and economic benefits in Australian regions, particularly through integrated growing, harvesting, transport and processing activities (Bureau of Rural Sciences 2005).

These issues are important as any expansion of the plantation estate will require access to rural land and engagement with landowners and communities. A related issue is the relationship between large-scale and small-scale plantation forestry where the latter are generally viewed more positively (Schirmer 2007). This has led to greater policy recognition of the need for improved relationships between the plantation industry and the rural sector to achieve an appropriate mix of industrial scale and small scale plantings (de Fegely et al. 2011, Ferguson 2014a). To this end, Gordon et al. (2013) assessed the extent of community engagement by the Australian plantation forest industry, and while stakeholders often considered the engagement to be ineffective, measures were identified to enhance relationships with stakeholders and the required skills of forestry professionals.

DISCUSSION

Internationally, the development of wood plantations has struggled to attract large-scale private investment without some form of government intervention or subsidy (Nielson 2007). This is primarily because of the time value of money with high up-front establishment costs and a long period until harvest returns. This, in turn, results in a lower rate of return for plantations compared with other investments excepting where there is other comparative advantage (e.g. high product prices, low transport costs, high yields). This has been the case in Australia, and to some extent New Zealand, where the high opportunity cost of land for agriculture has inhibited new investment without some form of government support. This is reflected in the cumulative plantation forest area in Australia over time (Figure 1), where early direct intervention (i.e. state ownership) and assistance schemes generated significant new plantings from the early 1960s.

However, it is important that development policies are well-targeted to address market failure, minimise distortions and promote the long-term competitiveness of plantations. This is particularly relevant in terms of the potential over-stimulation of the sector through previous subsidy experiences in New Zealand (Rhodes and Novis 2002), and more recent uptake of taxation measures in Australia. In Australia and New Zealand, government support has now transitioned to more indirect enabling policies, with a recent decline in new plantings partly as a consequence of market conditions (Figure 4).

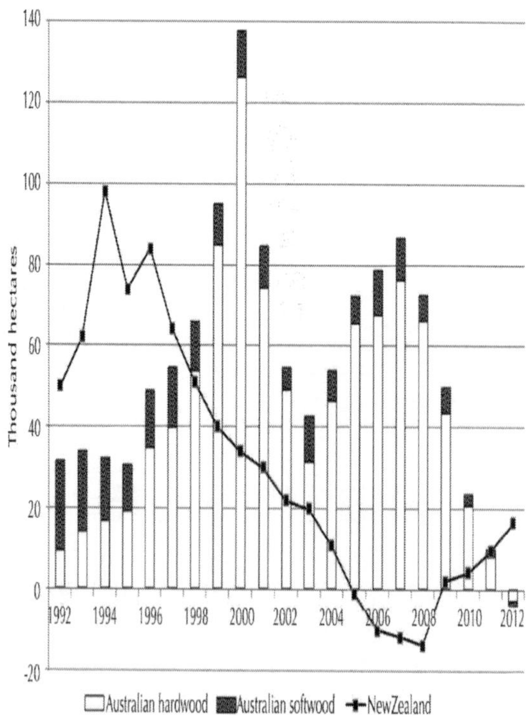

Figure 4: New plantations established in Australia and New Zealand (*Source*: ABARES and New Zealand Forest Owners Association. Note: new plantation areas are net of plantation conversion to other land use).

A key trend is the sharp decline in hardwood (i.e. eucalypt) planting rates since 2007 in Australia and plateauing or fall in softwood plantations in both Australia and New Zealand since the 1990s. This will act as a constraint to the processing industry's ambitions to expand the scale of processing to achieve internationally cost-competitive levels using local wood and fibre (de Fegely et al. 2011).

On the export side, plantation owners in Australia and New Zealand have taken advantage of increasing demand for both softwood and hardwood logs and woodchips in China. With respect to the eucalypt plantation resource, Australian producers have experienced a recent upswing in exports to China, although price competitiveness with other emerging regional suppliers such as Vietnam will be important. In June 2014, Vietnam took over from Australia as the world's largest supplier of woodchips to pulp mills in Asia (Pulp and Paper News 2014).

The outlook for plantation forest product exports to Asia from Australia and New Zealand will be influenced by the competitiveness of other major suppliers. The expansion of fast-growing wood plantations (i.e. rotations less than 40 years) in many parts of the world represents a challenge in terms of improved growth (measured as mean annual increment (MAI)) (Table 4) and increased supply. While growth rates in Australia and New Zealand are comparable to many other countries, there can be considerable variation as in the case of Australia (Table 5). Growth rates vary depending on tree species and region, reflecting underlying differences in climatic and edaphic conditions.

Table 4: Selected international growth rates for wood plantations

Species	**Mean Annual Increment** (m^3 ha^{-1} yr^{-1})	**Rotation length** (years)	**Main countries or regions**
Eucalyptus grandis W.Hill ex Maiden and various hybrids	15 - 40	5 - 15	Brazil, South Africa, Uruguay, India, Congo, Zimbabwe
Other tropical eucalypts	10 - 20	5 - 10	China, India, Thailand, Vietnam, Madagascar, Myanmar
Temperate eucalypts	5 - 18	10 - 15	Chile, Portugal, north-west Spain, Argentina, Uruguay, South Africa, Australia
Tropical acacias	15 - 30	7 - 10	Indonesia, China, Malaysia, Vietnam, India, Philippines, Thailand

Caribbean pines	8 - 20	10 - 18	Venezuela
Pinus patula Schiede ex Schltdl. & Cham. and*P. elliotii*	15 - 25	15 - 18	Swaziland
Gmelina arborea Roxb.	12 - 35	12 - 20	Costa Rica, Malaysia, Solomon Islands
Paraserianthes falcataria (L.) Nielsen	15 - 35	12 - 20	Indonesia, Malaysia, Philippines
Poplars	11 - 30	7 - 15	China, India, USA, central and western Europe, Turkey
Pinus radiata *	25 - 30	25	Australia, New Zealand, South Africa, Chile

Source: Cossalter and Pye-Smith (2003), * FAO (2001).

Rhodes and Stephens New Zealand Journal of Forestry Science 2014 44(Suppl 1):S10, doi:10.1186/1179-5395-44-S1-S10

Table 5: Australian regional growth rates for wood plantations

Species	**Mean Annual Increment** (**m^3 ha^{-1} yr^{-1}**)	**Rotation length** (**years**)	**Region(s)**
Temperate eucalypts (pulpwood)	10 - 19	12 - 25	Tasmania, South Australia, New South Wales, Victoria, Queensland, south-west Western Australia
Temperate eucalypts (sawlog)	20 - 27	14 - 20	Tasmania, South Australia, New South Wales, Victoria, Queensland, south-west Western Australia
Araucaria spp.	15	50	Queensland, northern New South Wales
Pinus radiata	16 - 21	30	Tasmania, Victoria, New South Wales, south-west Western Australia, South Australia
Pinus elliotii and *P. caribaea*hybrids	15 - 18	30	Queensland, northern New South Wales
Eucalyptus nitens	15	15	Tasmania

Source: Gavran et al. (2012).

Rhodes and Stephens New Zealand Journal of Forestry Science 2014 44(Suppl 1):S10, doi:10.1186/1179-5395-44-S1-S10

However, the long-term sustainability, expansion and productivity of many international wood plantations are uncertain, reflecting underlying environmental and social issues. These issues include population pressure, small land holdings, technology transfer, environmental practices and competition for land and food production (Cossalter and Pye-Smith 2003, Kugelman and Levenstein 2012). The Asia Pacific Forestry Commission (2010) noted that while plantation productivity may increase, this will depend in many countries on improving favourable policy and institutional environments. In the case of South East Asia, for example, Nambiar and Harwood (2014) identify considerable wood plantation potential, but this is dependent upon coordinated research and management for genetic improvement and soil protection.

By comparison, Australia and New Zealand have well developed institutional frameworks, research capacity and environmental policies for sustainable forest management and improved productivity. There has been over 100 years of plantation research and development focusing initially on exotic pines, and there is considerable scope to lift the productivity of the new eucalypt plantation estate in Australia. Genetic improvement and better site matching of species is expected to deliver higher growth for plantation eucalypts over the next rotation, given the rationalization of the former MIS estate in Australia. The New Zealand forest industry science and innovation strategy has also identified a 5 m^3 ha^{-1} $year^{-1}$ increase in productivity by 2025.

The other key issue is what type of wood plantation industry is feasible or desirable in Australia and New Zealand into the future? This is a broader policy issue, reflecting the economics of exporting relatively unprocessed industrial wood versus domestic value adding, which will have implications for future national policy, plantation management and investment. Ferguson (2014a)argues that the Australian plantation industry is at a cross-roads and that a coherent national plantation policy is needed if the Australian government and industry wish to expand the plantation estate to maintain international competitive wood and paper manufacturing and associated rural employment.

Manufacturing costs in Australia and New Zealand can be at the higher end of global benchmarks, factoring in existing scale of facilities, environmental standards, labour and energy. For these reasons, there has been considerable interest in research and innovation into emerging opportunities to value add either low grade logs or under-utilised biomass residues such as for bioenergy, as well as higher value applications for sawn timber and engineered wood products.

Bioenergy offers potential as a new revenue stream that is dependent on renewable energy policy as well as available feedstocks and technology developments (International Energy Agency 2009). The Clean Energy Council (2008) identified there was enough wood waste from forestry and processing activity in Australia to generate up to 3000 Gwh per year or 7% of the national renewable energy target. In New Zealand, over $NZ13 million has been invested into research into converting forestry biomass into liquid biofuels by looking into the commercial viability of establishing a modular test plant to process forest waste into sustainable transport fuel (Ministry for Primary Industries 2013b). These opportunities reflect the growing demand for renewable energy sources and provide a medium to longer term market for the plantation wood industry.

However, it will be equally important that some of the social challenges of plantation development be managed into the future, given previous community concerns over plantation expansion and related socio-economic and environmental impacts. The need for improved community engagement is considered important in gaining wider acceptance from stakeholders (Gordon et al. 2013). Plantation development can occur across a continuum of planting types from monocultures of exotic species to stands of mixed native species; and varying scales from broad scale plantations to small farm woodlots and riparian plantings. The majority of wood plantations in Australia and New Zealand have been large-scale and restricted to a select number of species for high productivity and commercial purposes. Community preferences for more diversified and small-scale plantings may entail changes in planting design and species mix which can have direct impacts on the costs to the forest owner. Managing plantations as part of whole farm plans may assist in this process, although would need to be weighed against the commercial costs and benefits such as higher transaction costs from more dispersed plantation areas.

This leads to the other recent trend of increasing interest in planted forests for ecosystem services. To date, it has been difficult to quantify the extent of these types of plantings as they are undertaken by a range of individuals such as farmers, community based organisations, private businesses, family trusts and local governments. In Australia, there is estimated to be 153 000 ha of non-industrial planted forest.

The outlook for ecosystem services plantings other than for carbon sequestration in Australia and New Zealand is uncertain but likely to increase in conjunction with direct Government programmes, albeit on a small scale, for the amelioration of land degradation. In many cases these programmes are oriented toward revegetation using mixed plantings of endemic native species or natural regeneration. The other major activity is farm forestry activities for wood production, farm productivity and environmental amenity. In these cases, forestry extension is important and investment decisions can be more complex involving lifestyle, income, cash flow and risk choices for the landowner (Schirmer et al. 2000).

In terms of the broad potential for carbon plantings, either as not-for-harvest or joint carbon and wood plantations, it has been shown that prices above \$20/t CO_2 could stimulate new plantation investment in Australia and New Zealand. However, a number of significant impediments are identified in the respective national climate change policy schemes. In the case of Australia and New Zealand, this includes a lack of inclusion of harvested wood products in applicable methods and recognised afforestation activities and in Australia a relatively short contract period for the purchase of emission reductions by the Australian Government. In the case of New Zealand, it is also the partial coverage of the scheme for emitters and uncapped link to international carbon credits that has depressed carbon prices to less than \$3/t CO_2 in recent years. Other issues have included risk and uncertainty through volatility in carbon prices, particularly until underlying national and international carbon regulatory frameworks are fully developed.

CONCLUSIONS

Planted forest development in Australia and New Zealand has followed a similar pattern with a strong historical focus on industrial wood plantations. The net outcome from these policies has been the development of a sizeable wood plantation resource and associated downstream processing. However, once governments have transitioned away from direct incentives to an enabling policy environment there has been little new plantation investment.

Concurrently, markets have emerged for some ecosystem services such as landscape restoration, biodiversity and carbon sequestration.

Further development and access to these markets may help address some of the investment hurdles for new wood plantations while delivering multiple benefits. There is therefore considerable interest in addressing market failure through the use of emerging markets for the joint production of wood and ecosystem services. A market based solution for ecosystem services in wood plantations has the added potential benefit of increasing economies of scale and competitiveness in wood resource availability and associated downstream processing.

It is argued that more coherent national plantation policies are needed to capture the full range of benefits from wood plantations and other planted forests. In 1997, the Plantations 2020 vision in Australia had a target of trebling the plantation area to 3 million hectares by 2020 to foster economic development. Similarly, the New Zealand government has a target of increasing the planted area by 20 000 hectares per annum, largely to meet climate policy objectives. These goals are unlikely to be met, at least in the short term, given falling rates of new plantings. In fact, there was a net decrease in Australia in the total area of wood plantations by 4 000 ha in 2011-12 (Gavran 2013). In New Zealand, recent surveys have identified deforestation intentions of many landowners, particularly conversion to dairy, over the next few years (Manley 2014). Better design of policy settings, particularly for carbon markets, and policy consistency between land uses, may curb falling rates of new plantings and encourage new plantings to help achieve climate change mitigation and economic development goals.

For existing wood plantations, growing demand in Asia, particularly in China and India, offers significant opportunities for export growth of round wood and woodchips from Australia and New Zealand. This outlook needs to be assessed against the growing supply of forest products from Latin America and Asia, the sustainability of which will depend on underlying policies for environmental management and institutional capacity. There is also potential for growth in bioenergy and other value added products from plantations in Australia and New Zealand over the medium term, either for domestic consumption or export.

In order to fully realise the opportunities for wood plantation development, enhanced community engagement will help address previous concerns regarding the acceptability of plantations in the rural landscape. Close attention will also need to be applied to managing future climate change risks, given forecast changes in temperature and rainfall and secondary effects such as increased pests and diseases or more extreme weather. Climate change forecasts present both risks and opportunities, through predicted increases or decreases in forest growth across regions and other risks posed by drought, fire and wind.

AUTHORS' CONTRIBUTIONS

David Rhodes and Michael Stephens jointly conceived the review as part of their participation in the Third International Congress on Planted Forests, contributing New Zealand and Australian planted forest perspectives respectively and underlying comparative analyses. Both authors reviewed the literature, drafted the thematic sections and approved the final manuscript.

ACKNOWLEDGEMENTS

The valuable suggestions made by two anonymous referees are gratefully acknowledged.

REFERENCES

1. Australian Bureau of Agricultural and Resource Economics and Sciences (2011). Potential effects of climate change on forests and forestry in Australia. Canberra, Australia: ABARES. August
2. Australian Bureau of Agricultural and Resource Economics and Sciences (2013). Preliminary long-term forecasts of wood product demand in Australia: research report 13.6. Canberra, Australia: ABARES. May
3. Australian Bureau of Agricultural and Resource Economics and Sciences (2014). Australian Forest and Wood Products Statistics: September and December quarters 2013. Retrieved 28 May 2014 from: http://www.daff.gov.au/ABARES/pages/publications/display.aspx?url=http://143.188.17.20/anrdl/DAFFService/display.php?fid=pb_afwpsd9abfe20140526_11a.xml

4. Australian Bureau of Statistics (2014). 3222.0 Population Projections, Australia, 2012 (base) to 2101. Retrieved 28 May 2014 from:http://www.abs.gov.au/ausstats/abs@.nsf/Lookup/3222.0main+features32012%20(b se)%20to%202101
5. Asia Pacific Forestry Commission (2010). Asia-Pacific forests and forestry to 2020. Report of the second Asia-Pacific forestry sector outlook study. Bangkok: Food and Agriculture Organization of the United Nations.
6. Bahaus, J, van der Meer, P, Kanninen, M (2010). Ecosystem goods and services from plantation forests. London/Washington: Earthscan.
7. Baral, H, Keenan, RJ, Fox, JC, Stork, NE, Kasel, S (2013). Spatial assessment of ecosystem goods and services in complex production landscapes: a case study from south-eastern Australia. *Ecological Complexity, 13*, 35–45.
8. Battaglia, M, Bruce, J, Brack, C, Baker, T (2009). Climate change and Australia's plantation estate: analysis of vulnerability and preliminary investigation of adaptation options. (FWPA) Project No. PNC068-0709). Melbourne: Forest and Wood Products Australia/CSIRO.
9. Booth, TH, Kirschbaum, MUF, Battaglia, M In C. Stokes & M Howden (Ed.) (2010). Forestry.*Adapting agriculture to climate change, Chapter 9*, 137–152. Victoria, Australia: CSIRO Publishing.
10. Buongiorno, J, Zhu, S, Raunikar, R, Prestemon, JP (2012). Outlook to 2060 for world forests and forest industries: a technical document supporting the Forest Service 2010 RPA assessment. Asheville, North Carolina: United States Department of Agriculture, Southern Research Station.
11. Bureau of Rural Sciences (2005). Socioeconomic impacts of plantation forestry. Melbourne, Australia: Forest and Wood Products Research and Development Corporation. November
12. Carle, J, & Holgrem, P (2008). Wood from planted forests: a global outlook 2005-2030. *Forest Products Journal, 58*, 6–18.
13. Clean Energy Council (2008). Australian bioenergy roadmap: setting the direction for biomass in stationary energy to 2020 and beyond. Melbourne, Australia. September

14. Commonwealth of Australia (2013). Emissions Reduction Fund Green Paper. Canberra: Department of the Environment.
15. Cossalter, C, & Pye-Smith, C (2003). Fast-wood forestry: myths and realities. Forest Perspectives. Bogor: Center for International Forestry Research, Indonesia.
16. Dale, G, & Dieters, M (2007). Economic returns from environmental problems: breeding salt- and drought- tolerant eucalypts for salinity abatement and commercial forestry. *Ecological Engineering,31*, 175–182.
17. de Fegely, R, Stephens, M, Hansard, A (2011). Review of policies and investment models to support continued plantation investment in Australia. (Project No. PRA189-1011). Melbourne, Australia: Forest and Wood Products Australia. March
18. Dickinson, G, Reilly, D, Lindsay, DAJ (2011). Abstracts from Darwin 2011 - African Mahogany Plantation Industry Forum, Darwin, 30 August - 1 September 2011. Brisbane, Australia: Queensland Department of Employment, Economic Development and Innovation.
19. Fairweather, J, & Swaffield, S (2003). Public perceptions of natural character and implications for the forest sector. *New Zealand Journal of Forestry, 47*, 24–30.
20. Ferguson, I (2014a). Australian plantations: mixed signals ahead. *International Forestry Review,16*, 160–171.
21. Ferguson, I (2014b). Australian plantation inventory: ownership changes, availability and policy.*Australian Forestry, 77*, 25–38. Publisher Full Text
22. Food and Agriculture Organization (FAO) (2001). Mean annual volume increment of selected industrial forest plantation species. Forest Plantation Thematic Papers, Working Paper 1. Rome: Food and Agriculture Organization of the United Nations. April
23. Food and Agriculture Organization (FAO) (2006a). Global forest resource assessment 2005. (FAO Forestry Paper 147). Rome: Food and Agriculture Organization of the United Nations.
24. Food and Agriculture Organization (FAO) (2006b). Global planted forests thematic study: results and analysis. (Planted Forests and Trees Working Paper 38). Rome: Food and Agriculture Organization of the United Nations.

25. Food and Agriculture Organization (FAO) (2010). Global forest resources assessment 2010, main report. (FAO Forestry Paper 163). Rome: Food and Agriculture Organization of the United Nations.
26. Gavran, M, Frakes, I, Davey, S, Mahendrarajah, S (2012). Australia's plantation log supply 2010-2054. Canberra: Australian Bureau of Agricultural and Resource Economics and Sciences. May
27. Gavran, M (2013). Australian plantation statistics 2013 update. (Technical report 13.3). Canberra: Australian Bureau of Agricultural and Resource Economics and Sciences. May
28. Global Construction Perspectives and Oxford Economics (2013). Global construction market to reach $15 trillion by 2025. Retrieved 12 March 2014 from: http://www.globalconstruction2025.com/media/wysiwyg/press-release/GCP-2025-press-release.pdf *webcite*
29. Gordon, M, Schirmer, J, Lockwood, M, Vanclay, F, Hanson, D (2013). Being good neighbours: current practices, barriers, and opportunities for community engagement in Australian plantation forestry. *Land Use Policy, 34*, 62–71.
30. Hale, T, Kahui, V, Farhat, D (2014). A modified production possibility frontier for efficient forestry management under the New Zealand Emissions Trading Scheme. *Australian Journal of Agricultural and Resource Economics, 58*, 1–17.
31. Hock, B, Payn, T, Clinton, P, Turner, J (2009). Towards green markets for New Zealand plantations. *New Zealand Journal of Forestry, 54*, 9–19.
32. Hennessy, K, Fitzharris, B, Bates, BC, Harvey, N, Howden, SM, Hughes, L, Salinger, J, Warrick, R In ML Parry, OF Canziani, JP Palutikof, PJ van der Linden & CE Hanson (Eds.) (2007). Australia and New Zealand. *Climate Change 2007: impacts, adaptation and vulnerability Contribution of Working Group II to the Fourth Assessment Report of the Intergovernmental Panel on Climate Change* (pp. 507–540). Cambridge, UK: Cambridge University Press.
33. Indufor (2012). Strategic review on the future of forest plantations. Retrieved 12 March 2014 from: http://ic.fsc.org/download.strategic-review-on-the-future-of-forest-plantations-full-report.672.htm

34. International Energy Agency (2009). Bioenergy - a sustainable and reliable energy source: a review of status and prospects. (IEA Bioenergy: ExCo 2009:06, Main Report). Paris: International Energy Agency.
35. International Energy Agency (2011). Global wood pellet industry market and trade study. (IEA Bioenergy Task 40: sustainable international bioenergy trade). Paris: International Energy Agency. December
36. Kavanaugh, RP, Stanton, MA, Herring, MW (2007). Eucalypt plantings on farms benefit woodland birds in south eastern Australia. *Austral Ecology, 32*, 635–650.
37. Keenan, RJ, Davey, S, Grieve, A, Moran, B, Donaldson, J In HC Sim, S Appanah, & YC Youn (Eds.) (2004). Market mechanisms and assessment methods for environmental services from private forests in Australia. *Proceedings of the workshop Forests for poverty reduction: opportunities with Clean Development Mechanism, environmental services and biodiversity* Bangkok: Food and Agriculture Organization.
38. Kugelman, M, & Levenstein, SL (2012). The global farms race: land grabs, agricultural investment, and the scramble for food security. Washington DC: Island Press.
39. Lamers, P, Junginger, M, Hamelinck, C, Faaij, A (2012). Developments in international solid biofuel trade - an analysis of volumes, policies and market factors. *Renewable and Sustainable Energy Reviews, 16*, 3176–3199.
40. Langer, L, & Barnard, T (2002). Local community attitudes to plantation forestry, Gisborne/East Coast region. *New Zealand Journal of Forestry, 47*, 14–16.
41. Lindenmayer, DB, & Hobbs, RJ (2004). Fauna conservation in Australian plantation forests - a review. *Biological Conservation, 119*, 151–168.
42. Lippke, B, Oneil, E, Harrison, R, Skog, K, Gustavvson, L, Sathre, R (2011). Life cycle impacts of forest management and wood utilization on carbon mitigation: knowns and unknowns. *Carbon Management, 2*, 303–333.
43. Maclaren, JP (1996). Environmental effects of planted forests in New Zealand: the implications of continued afforestation of

pasture. (Forest Research Institute Bulletin No. 198). Rotorua, New Zealand: New Zealand Forest Research Institute.

44. Malsheimer, RW, Bowyer, JL, Fried, JS, Gee, E, Izlar, R, Miner, RA, Munn, IA, Oneil, E, Stewart, WC (2011). Managing forests because carbon matters: integrating energy, products and land management policy. *Journal of Forestry, 109*(7S), S7–S50.

45. Manley, B, & Maclaren, P (2009). Modelling the impact of carbon trading legislation on New Zealand's plantation estate. *54*, 39–44. New Zealand Journal of Forestry.

46. Manley, B (2014). Deforestation Survey 2013 Final Report. Ministry for Primary Industries Technical Paper 2014/07, February.

47. McDermott, CL, Cashore, B, Kanowski, P (2010). Global environmental forest policies: an international comparison. (Chapter 8: Oceania: Australia and New Zealand). London, New York: Earthscan.

48. Ministry for Primary Industries (2013a). Situation and outlook for primary industries 2013, June. Retrieved 29 August 2014 from: http://www.mpi.govt.nz/news-resources/news/situation-amp-outlook-for-primary-industries

49. Ministry for Primary Industries (2013b). Stump to Pump programme receives PGP funding boost. [Media release, 25 July 2013]. Retrieved 10 June 2014 from: http://www.mpi.govt.nz/news-resources/news/stump-to-pump-programme-receives-pgp-funding-boost

50. Ministry for Primary Industries (2014). Forestry statistics. Retrieved 29 August 2014 from: http://www.mpi.govt.nz/forestry/statistics-forecasting/forestry-statistics

51. Mitchell, CD, Harper, RJ, Keenan, RJ (2012). Current status and future prospects for carbon forestry in Australia. *Australian Forestry, 75*, 200–212.

52. Montreal Process Implementation Group for Australia and National Forest Inventory Steering Committee (2013). Australia's State of the Forests Report 2013. Canberra: ABARES. December

53. Nambiar, S, & Ferguson, I (2005). New forests: wood production and environmental services. Melbourne, Australia: CSIRO Publishing.

54. Nambiar, EKS, & Harwood, CE (2014). Productivity of acacia and eucalypt plantations in South-east Asia. 1. Bio-physical determinants of production: opportunities and challenges. *International Forestry Review, 16*, 225–248.
55. National Land and Water Resources Audit (2001). Australian dryland salinity assessment 2000: extent, impacts, processes, monitoring and management options. Canberra: National Land and Water Resources Audit.
56. New Zealand Forest Owners Association (2013). New Zealand plantation forest industry. Facts and figures. Retrieved 14 April 2014 from: http://www.nzfoa.org.nz/publications-5/facts-and-figures
57. Nielson, DA (2007). Corporate private sector dimensions in planted forest investments. (Planted Forests and Trees Working Paper Series FP/40E). Rome: Food and Agriculture Organization of the United Nations.
58. Oliver, CD, Nassar, NT, Lippke, BR, McCarter, JB (2014). Carbon, fossil fuel, and biodiversity mitigation with wood and forests. *Journal of Sustainable Forestry, 33*, 248–275.
59. Paul, KI, Reeson, A, Polglase, PJ, Ritson, P (2013). Economic and employment implications of a carbon market for industrial plantation forestry. *Land Use Policy, 30*, 528–540.
60. Pulp and Paper News (2014). Vietnam now world's largest woodchip supplier. Retrieved 9 June 2014 from: http://www.pulpapernews.com/2012/03/vietnam-now-world-s-largest-wood-chip-supplier
61. Rhodes, D (2001). Rehabilitation of deforested steep slopes on the east coast of New Zealand's North Island. *Unasylva, 207*, 21–29.
62. Rhodes, D, & Novis, J (2002). The impact of incentives on the development of plantation forest resources in New Zealand. (MAF Information Paper No. 45). Wellington, New Zealand: Ministry of Agriculture and Forestry.
63. Schirmer, J (2007). Plantations and social conflict: exploring the differences between small-scale and large-scale plantation forestry. *Small-scale Forestry, 6*, 19–33.

64. Schirmer, J, Kanowski, P, Race, D (2000). Factors affecting adoption of plantation forestry on farms: implications for farm forestry in Australia. *Australian Forestry, 63*, 44–51.
65. Singh, S, Davey, S, Cole, M (2010). Implications of climate change for forests, vegetation and carbon in Australia. *New Zealand Journal of Forestry Science, 40*, 141–152.
66. Sochacki, SJ, Harper, RJ, Smettem, KRJ (2012). Bio-mitigation of carbon following afforestation of abandoned salinized farmland. *Global Change Biology (GCB) Bioenergy, 4*, 193–201.
67. State of the Environment Committee (2011). Australia state of the environment 2011. (Independent report to the Australian Government Minister for Sustainability, Environment, Water, Population and Communities). Canberra.
68. Stephens, M (2010). Bushfire, forests and land management policy under a changing climate. *Farm Policy Journal, 7*, 11–19.
69. Stephens, M, Pinkard, L, Keenan, R (2012). Plantation forest industry climate change adaptation handbook. Canberra: Australian Forest Products Association. [http://www.ausfpa.com.au/wp-content/uploads/Change-Adaptation-Handbook.pdf]
70. Stephens, ML, & Grist, P (2014). Market failure for plantations: past experiences and emerging trends for delivering wood production and ecosystem services in Australia. *International Forestry Review, 16*, 205–215.
71. Townsend, PV, Harper, RJ, Brennan, PD, Dean, C, Wu, S, Smettem, KRJ, Cook, SE (2012). Multiple environmental services as an opportunity for watershed restoration. *Forest Policy and Economics,17*, 45–58.
72. URS Forestry (2008). Farm forestry area and resources in Australia. (RIRDC Publication No. 08/104). Canberra: Rural Industries Research and Development Corporation.
73. Watt, MS, Kirschbaum, MUF, Paul, TSH, Tait, A, Pearce, HG, Brockerhoff, EG, Moore, JR, Bulman, LS, Kriticos, DJ (2008). The effect of climate change on New Zealand's planted forests: impacts, risks and opportunities. (Ministry of Agriculture and Forestry Report No. CC MAF POL_2008-07 (106-1) No. 1, July). Wellington, New Zealand.

74. West, G, Wakelin, S, Wall, A, Turner, J, Poole, B (2011). Achieving multiple regional goals with carbon forestry. (A Scion report for Waikato Regional Council). Rotorua, New Zealand: Scion. [http://www.waikatoregion.govt.nz/PageFiles/19232/scion%20 report.pdf] *webcite*
75. Williams, K (2008). Community attitudes to plantation forestry. (Technical Report 194). Cooperative Research Centre for Forestry and Department of Resource Management and Geography, University of Melbourne.
76. Wilson, S, Bhati, UN, Stephens, M (1995). Trees on Australian broadacre and dairy farms. Quarterly forest products statistics: September quarter. Canberra: Australian Bureau of Agricultural and Resource Economics.
77. Wood Council of New Zealand (2012). New Zealand forest and wood products industry strategic action plan. Wellington, New Zealand. Retrieved 12 March 2014 from:http://www.woodco.org.nz/images/stories/pdfs/ForestWood_Strategic_Action_Plan.pdf
78. Yao, RT, Scarpa, R, Turner, JA, Barnard, TD, Rose, JM, Palma, JHN, Harrison, DR (2014). Valuing biodiversity enhancement in New Zealand's planted forests: socioeconomic and spatial determinants of willingness-to-pay. *Ecological Economics, 98,* 90–101.
79. Zhang, L, Dowling, T, Hocking, M, Morris, J, Adams, G, Hickel, K, Best, A, Vertessy, R (2003). Predicting the effects of large-scale afforestation on annual flow regime and water allocation: an example for the Goulburn-Broken catchments. (Technical Report 03/5). Canberra: CRC for Catchment Hydrology.
80. Zhang, L, Vertessy, R, Walker, G, Gilfedder, M, Hairsine, P (2007). Afforestation in a catchment context: understanding the impacts on water yield and salinity. (Industry report 1/07). Melbourne, Australia: eWater CRC.

Citations

CHAPTER 1

Jens Keilwagen, Benjamin Kilian, Hakan **Özkan**, Steve Babben, Dragan Perovic, Klaus F. X. Mayer, Alexander Walther, C. Hart Poskar, Frank Ordon, Kellye Eversole, Andreas Börner,Martin Ganal, Helmut Knüpffer, Andreas Graner, and Swetlana Friedel, Separating the Wheat from the Chaff – a Strategy to Utilize Plant Genetic Resources from *Ex Situ*genebanks, doi:10.1038/srep05231.

CHAPTER 2

Yizhen Wan, Heidi R Schwaninger, Angela M Baldo, Joanne A Labate, Gan-Yuan Zhong, and Charles J Simon, A Phylogenetic Analysis of the Grape Genus (*Vitis* L.) Reveals Broad Reticulation and Concurrent Diversification during Neogene and Quaternary Climate Change, doi: 10.1186/1471-2148-13-141.

CHAPTER 3

Wolfgang Falk and Nils Hempelmann, "Species Favourability Shift in Europe due to Climate Change: A Case Study for Fagus sylvatica L. and Picea abies (L.) Karst. Based on an Ensemble of Climate Models," Journal of Climatology, vol. 2013, Article ID 787250, 18 pages, 2013. doi:10.1155/2013/787250.

CHAPTER 4

Madhav Pandey and Om P Rajora, Higher Fine-Scale Genetic Structure in Peripheral than In Core Populations of a Long-Lived and Mixed-Mating Conifer - Eastern White Cedar (Thuja Occidentalis L.), doi:10.1186/1471-2148-12-48.

CHAPTER 5

Gulab D Khedkar, Rahul Jamdade, Amol Kalyankar, Anita Tiknaik, Tetsuzan Benny Ron, and David Haymer, Genetic fragmentation in India's third longest river system, the Narmada, doi:10.1186/2193-1801-3-385.

CHAPTER 6

Kyoko Okano and M Syndonia Bret-Harte, Warming and Neighbor Removal Affect White Spruce Seedling Growth Differently Above and Below Treeline, doi:10.1186/s40064-015-0833-x.

CHAPTER 7

Nourollah Ahmadi, Alain Audebert, Malcolm J Bennett, Anthony Bishopp, Antonio Costa de Oliveira, Brigitte Courtois, Abdala Diedhiou, Anne Diévart, Pascal Gantet,Alain Ghesquière, Emmanuel Guiderdoni, Amelia Henry, Yoshiaki Inukai, Leon Kochian, Laurent Laplaze, Mikael Lucas, Doan Trung Luu, Baboucarr Manneh,Xiaorong Mo, Raveendran Muthurajan, Christophe Périn, Adam Price, Sabariappan Robin, Hervé Sentenac, Bassirou Sine, Yusaku Uga, Anne Aliénor Véry, Matthias Wissuwa, Ping Wu, and Jian Xu, The Roots of Future Rice Harvests, doi:10.1186/s12284-014-0029-y.

CHAPTER 8

David Rhodes and Michael Stephens, Planted Forest Development in Australia and New Zealand: Comparative Trends and Future Opportunities, doi:10.1186/1179-5395-44-S S10.

Index